Previous Works by William Moreira (Canno)

Life Is Beautiful, Doesn't Matter What, because We Are God's Children
Embraced in Love, We Reach Heaven
(220 pages)

How to Suffer Happily
Dr. Fritz, the Phenomenon of the Millennium (440 pages)

What Christ, Thomas Paine, and Allan Kardec
Want You to Know and Religion Doesn't
(220 pages)

Christ's Wisdom and the Unholy Prophets
(640 pages)

GOD! The Realities of the Creator
(260 pages)

THE BIG NEST
Originated the
BIG BANG
of Stephen Hawking's Black Holes

HOPE: THE ANSWER TO THE NIHILISM OF MODERN PHYSICS

William Moreira

Sixty-one years of journalism and writing

(eight books published)

Life Sciences/Philosophy/Realities/Spiritualism

iUniverse LLC
Bloomington

THE BIG NEST ORIGINATED THE BIG BANG OF STEPHEN HAWKING'S BLACK HOLES
HOPE: THE ANSWER TO NIHILISM OF MODERN PHYSICS

iUniverse books may be ordered through booksellers or by contacting:

iUniverse
1663 Liberty Drive
Bloomington, IN 47403
www.iuniverse.com
1-800-Authors (1-800-288-4677)

ISBN: 978-1-4759-9676-0 (sc)
ISBN: 978-1-4759-9677-7 (hc)
ISBN: 978-1-4759-9678-4 (e)

Library of Congress Control Number: 2013911433

Printed in the United States of America.

iUniverse rev. date: 8/23/2013

In Memory

Carol Jacqueline Moreira (1962–1996)

Nothing reaches as deeply into our minds as moral pain, as the tragic death of our child, especially after thirty-three marvelous years.

I cried for two weeks and couldn't pilot my Cessna for six months, and then one morning, as I was staring at the clouds from her bedroom window with a hollow heart because she wasn't there but maybe up there, I heard her voice loud and clear:

"Pappy, life is beautiful. It doesn't matter what, because we are God's children."

Was it a daydream or maybe just my mind in search of hope as a consolation? Was it God's mercifulness?

If it wasn't for the concept of a Creator as perfect as the universe, because there is no other way of imagining it, life wouldn't be worth one minute of breathing.

That night, I reached as high as my wings could take me, and there I said my prayers, this time to him and to her. I knew both were in the celestial dimensions where we will all be, because when love is involved, there can never be an eternal separation.

William Moreira (Canno)—a faithful father

Washington (DC) Author doing a interview for the Brazilian Consulate (35-years old) **1968**

New York City (NY) with wife Gladys and CAROL as a baby (22 years-old) **1955**

Palm Springs (CA-) his plane CESSNA 172 RG (65 years-old) **1998**

Lincoln Park (NJ) his plane
CESSNA 182-RG SKYLINE (64 years-old) **1997**.

PRESS CARD, Brazil)
(23 years-old) **1956**

JOURNALIST ID (Brazil)
20 years-old - **1953**

Contents

Introduction

I am now in my eighties. I have been an author, journalist, entrepreneur, pilot, chef, painter, designer, and everything under the stars; I am fluent in English, Portuguese, and Spanish and speak a little of a few more languages. I have visited dozens of countries and lived more than forty years in the United States. I have come to know and understand the people and their polemics, as I admired this great nation, which received me and millions of others from other countries with open arms.

I have read thousands of books and magazines; watched hundreds of hours of documentaries; and visited temples, churches, synagogues, mosques, religious meetings, and even voodoo ceremonies. I have eaten bread prepared by angels and demons to know what life means, because to live is to fight as souls and as spirits.

Whether religious or atheist, all feel like they are in a dark tunnel without an exit sign or a labyrinth where all the ways end in the same circle of mysteries without solutions. Life in the flesh extinguishes with the beginning of another life in the spirit, while we look at an empty pedestal.

I direct you to a light we all seek called *hope*, while I defend God against the incredulous who do not see or feel that an intelligent universe could not have come from the null or void—or better yet, a nothingness—because behind every grain is another grain, filling the vacuum or emptiness.

The scripture "And God created the heavens" gives us the concept of an Intelligence, which we can think of as an existence greater than ours, being before the universe. We will call it a *Mother-Universe*, or the *Total*, as the illumined *dark matter* existing as time and eternal matter with no beginning or ending, representing the eternal existence.

All my life, since I was a child, I have noticed the millions of destitute people in our society and all over the planet, who live as slaves, victims, and beggars in absolute misery, continuously being born and dying on the streets. They are the majority, who work, earning a minimum, starving salary, even in countries like Brazil, which they say is now doing well financially, in contradiction to the Favelas (the ghettos where slaves lived not long ago as ghosts in absolute misery), while the welfare available is just a loaf of bread compared to that of the first-world countries.

While "love and charity" are not found in every heart, the Creator has one eye on our good deeds and the other eye on the "School of Learning for Spirits" called earth. The lessons come from the physical and moral pain, which make us scream when something is wrong in our conscience: "Oh! God! Have mercy on our souls!"

Stephen William Hawking in his 2010 book *The Grand Designer* continues to scream at the four winds that *God did not create the universe*, or worse yet, there is no need for a God, as the universe created itself and was nothing before it came into existence, as all the beauty of the creation just popped up from nowhere. He is paying the price.

What is beyond my comprehension is why he is acclaimed as a "brilliant mind" and a "super genius." Even though mummified in his body and living beyond explanation, he says he is not cursed or paying for sins, as religion says; he says the phenomenon of his condition is the law of nature existing since it created itself. I hope he sees me for a surprise.

On the Internet, someone said, "God didn't take Hawking yet because he doesn't know what to do with him," but he's already doing it, allowing Hawking to be here as a statue, feeling the marvelousness of life on earth while sitting on the "divine nail," as a representation and not a punishment, because of his big bang—I mean his *big mouth*.

What has astonished me after eighty years of researching continually in perfect health and physically considered to have a great body, even now at my age, because science says there is no effect without a cause, is that fifty years ago, I promised Hawking that he and I would survive for him to pay the price of defamation and ingratitude publically for his unmerciful act of having transformed his glory by trying hard to take hope from all of us.

Life in our carnal bodies is a beauty. Now Hawking has said he wants to keep living as long he can push it; this is a lesson to all of us, as even being in a crippled body with all kinds of pain and death are parts of being here. Our bodies come from a fusion of a sperm and an ovum, just like the meat from irrational animals. We are also in this group, but rational, having not only the instinct for survival but the intelligence to look at the infinity of the illuminated dark matter encrusted within the stars and, in exaltation, face such a background to know there is a super power and he is not far away but everywhere as God.

Our existence is perfect, as our hearts continue to beat. Mine is doing well after eighty years, waiting for my contest with Hawking. I am not sitting on a "nail" but enjoying what is offered to us, including piloting as if I could go to meet him. I have survived for this book, telling my brothers and sister that "Satan" comes authorized by the Creator to test our intelligence and love.

I am now facing Hawking with his glory-less life, because anyone would trade all the material glory and fame to be able to walk and enjoy a thick, juicy rare cheeseburger with crisp fries at the local tavern and the fun of happy souls singing in the company of our companions and then to walk home looking at a sky embedded with stars on a night of loving. We could change the set and go to a fancy place, have a gourmet dinner, choose from an array of gastronomically delightful dishes while listening to Beethoven's celestial music from heaven, while we stare at the skylight dome with the crystal clear image of our infinity compact with stars, while the irrational animals have their noses in the grass seeking it by the instinct of survival.

When we arrive at home, we begin the blessed ritual of love that unites man and woman. Life begins with the act to keep us together as a family, but first, we make a brief prayer of thankfulness for having been created and for an eternal spiritual world on time and ahead of it, as we all will meet again if love is involved.

At the age of sixty-four, I earned my right to fly as a pilot, and I immediately went up to fifteen thousand feet on a clear winter night above the mountains of Pennsylvania to see God, and I did. Tears rolled down my face, washing all the negativity from Hawking's funny quotes comparing us to monkeys in evolution, as disconsolation without mercy to the suffering crowd. I wished he would seek the Supreme Intelligence, not in a eighteen-inch computer monitor but right alongside me where the view was more spiritual than the most accurate telescope humans have ever made. I felt myself being sucked into the *clear, illuminated dark matter,* the spoken paradise where everyone wants to go as spirits.

The feeling is we can penetrate into the marvelousness of the *crystal clear, illuminated dark matter* as it exists eternally as the dwelling place of our Creator. I know it is enough for me and you; as for the others, whom I call the members of *Hawking's Atheist Club,* they will be carrying their heavy cross of negativity through a bitter existence, but there is no hurry, as time doesn't exist because of eternity. It just keeps ticking continually, having the past in our memory bank and the future ahead as a surprise in our free will.

William Moreira (Canno)
January, 13, 2013
Copacabana
Rio de Janeiro

A Note from the Author

On December 13, 2009, my book of 250 pages, *God! The Realities of the Creator,* was born on the thirteenth top deck of the ill-fated *Costa Concordia.*

After thirteen months of the book's writing, on January 13, 2012, the great ocean liner *Concordia* sank. My passports have the thirteenth as the date. In another book, *Dr. Fritz, The Phenomenon of the Millennium*, the five times I went to Brazil while writing the book I arrived on the thirteenth, and I didn't plan it. I am writing this book while sailing on the *Grand Holiday of Iberia*, from Rio de Janeiro to Buenos Aires, for ten days, leaving the first of January and arriving on the tenth, but my computer arrived on the thirteenth for my final rewriting of the *Big Nest* to become a reality.

Now, I have one eye on New Year's of the year 2013, which may be our whole hope for living in peace or in our own hell.

But don't worry, the cruiser *Grand Holiday,* will not sink, because the captain is not an alcoholic but a family man, always looking up for angels.

To finalize this book and give myself a deserved break after almost twenty hours on the keyboard, rewriting and translating from Portuguese, I used the service of *Costa Favolosa* for seven days to be nowhere near the Brazilian coast and to remember the fabulous *Concordia*, now a ghost as a starship cruising among the galaxies. Then, I wrote a few more lines while enjoying the greatness of our world, as we deserve it or nothing would be created, as the ill-fated cruiser was my stage for my great book published thirteen months after its last trip.

1 Imagine!

Imagine a senior, somewhere around his eighties, jovial, looking good in appearance—no wrinkles, an athletic body, rosy skin—all smiles, and charismatic with a celestial voice. He appeared suddenly, as if he had materialized from the mist straight to me, while I was running along the large sidewalk at the famous Copacabana Beach in the wee hours of a cold July winter.

Immediately, I noticed he could literally explain the mysteries of life here and beyond, which have us stressed and even paranoid in our ignorance, able to see only what is in front of our noses, confused by the complexities and feeling as if we are in a dark tunnel or a labyrinth where all the exits open to the same circle: we are born, suffer, and die, as the end of the human journey, or go on to a mysterious and tenebrous unknowing beyond the grave, maybe as a spirit, in another endless tunnel of darkness of infinity where there is no light.

The gracious senior kept talking, and I felt ecstasy in listening to a stranger who had suddenly appeared in the dawn at the edge of the beachfront. I sat on a stone bench so he could better educate me. There was not a single soul around besides those in the fast-moving cabs.

He was dressed in white shorts, a light-pink shirt, and sandals; he had curly, white hair, a short beard, and a mustache, which gave him a peaceful, celestial charm, making me behave like a child confronting such spirituality.

He kept explaining, and I listened to what I was seeking but did not find in religion, science, philosophy, physiology, or theologies based on factors in our carnal world, where life is a challenge from the crib to the tumulus and at all times between.

This extraordinary senior affirmed his knowledge came as he educated himself by reading thousands of books; he was born with the pen in his hand and had been writing and designing since he was a child, doing everything under the stars. He did everything and felt the good and the bad as lessons in the school of life, as a pupil in a discarded human body, where the challenges are never-ending, but necessary for the spiritual evolution of the soul as a spirit, which would continue to a world beyond, not mysterious but glorious, a true life in our universe, created for us to live in it, enjoying and respecting one another, because there is only one Creator.

Into his eighties, he continues. I listened with humbleness. He had survived families, friends, and strangers, experiencing the horror of World War II, revolutions, dictators, kings and queens, catastrophes, and near-fatal accidents on land and in the air; he had wept tears of joy and sorrow in the mysteries of "why life is like this."

Now, in the cold and fog of the wee hours, sitting on a hard stone bench at the beach, he explained to me as a stranger old enough to be perhaps my grandfather, but knowing about the phenomena considered mysterious by religions and science, its research without results on the complexities of our existence and life beyond death, as "to be or not to be," as "to be born or not to be born," as another sequence in existence, proving an imaginary principle of 259 degrees in the 360-degree dogma of life as intelligent human beings, in a fleshy body, like the carnal, irrational ones, who have not intelligence but instincts of survival.

We as humans have instincts and intelligence, along with the free will, as rational animals giving us the power of imagination to analyze, create, judge, condemn, and decide whether there is or is not a Supreme Intelligence, which created and continues creating existence from the nothingness, an infinite mass of a lightless matter, as the base of everything, an inexistence but existent. Our scientists argued that from the null and void nothing will be possible, contradicting one professor of cosmology, Stephen Hawking, knowing the one frozen in the wheelchair asserting there is no Creator is making a demagogue of himself.

More than ever, as proved by the great telescopes, there are celestial explosions, where there is no beginning or end, like an endless aquarium without walls, on which is everything that exists, including our souls.

The atheists feel lost confronting a perfect grandiosity and intermingled with their incapacity of rationalizing the sufferings of the carnal body, which is predestined to end in the final act in the grave. They question the unquestionable Creator, which is silence and occult, but they overlook that we are the created and not the Creator; the master of the creation decided to look and take notes, rather than appear and take sides.

The lack of answers and contact with the spiritual world (which is only affirmed by few, who are unable to prove it) have human beings unconvinced about life as the existence and the total, where the next step could be our last eternally. Now, Hawking is commenting he wants to live, because his world is only what his nose can point out.

It brings him the senseless feelings of a life of anxiety and paranoia; the horizon only brings tempests or catastrophes in a planet where the oceans cover almost all the area, but we can die of thirst. A microscopic world has invisible enemies, and the aging process begins on the act of being born and guarantees our voyage back to where we came from as the eternal unknowing and as a sure end of life materially. The only option is a spiritual one, as imaginable as the Creator himself.

"But, sir, what do we do? Everything conspires against our lives. What could be our defense? It is inexistent, because in the past, there existed great civilizations with extraordinary people living in splendor like an anthill, but now their glory is just dust swirling around in the whispering wind, and we will be next, because time doesn't stop; it is the base for the future."

I took a deep breath and held it, to better listen and find out if there was a solution, which we all seek as intelligent beings, to the mystery of the beginning of everything. Theories are related to conjectures, suppositions, hypotheses, and everything else that points to ground zero. Theories are endless, but their results are like children's play. Nothing comes from them.

I followed my guide, questioning him, and he kept answering, like an encyclopedia from the outer dimensions.

"The opinions are so diverse in religions or sciences and include now philosophy, as millions with their opinions and frustrations in all fields are lost, because we are in a world restricted by the flesh, which limits the spirit's vision."

I decided to ask the senior my most important and final questions, such as: "Who is God? How did everything began? What is his origin?"

Those questions are the same for everyone and are in grasp of themselves. They are not in a tree trunk of wisdom but at its roots. I blinked, and the genius senior had mysteriously disappeared, just as he had appeared, an angel. He flew away as real, leaving me perplexed in a reality, as if in a great dream.

We were the only two for four miles of a curvy, sandy beach, with its wide sidewalk and massive waves pounding mercilessly on the shore. The early sunrise, with its striking red-and-yellow design of a glorious day forced me to run, seeking the marvelous ghost of the galaxies, but it was in vain. Well, my friends, it seems complicated, but it isn't. In the future, souls and spirits will understand many mysteries as realities, because everyone is a fragment of the creation.

The "school" we are in is called earth. Everyone will learn what is offered to us. Some will do it today and others tomorrow, but the learning is for all of us. It is mandatory, just like when I was eighteen and I had to serve in the army and answer the call of duty.

To confirm, as my eyewitness, I just had to look up. He is there, tall in his thirty-eight meters (114 feet) as the tallest statue on our planet, the Christ of Nazareth, with eyes and arms open. He is not crucified. Carved in white granite, he stands at the top of the heavens, a rock 703 meters high (2,109 feet) nested in the middle of Rio de Janeiro. Seeing, night and day up to fifty miles, he blesses the world and the universe, rated the Seventh Marvelous Statue by a unanimous international vote.

Christ the Redeemer was voted by all religions as an icon of love and not a religious issue because he gave us the hope of an eternal life where our existence will continue in spirit. Heaven has no limits and is free of the black holes and menaces of Hawking, for whom there is no hope of an existence beyond our material life.

The last question without answer is whether we will live eternally in hell or heaven. That will depend on your religious beliefs, because we cannot reach anything before the big bang, or better yet, the egg we came from as a fragment—not as a chicken but as a chick—where the universe began and ended at the gate of the fence. Before the nest and beyond the gate is the Universe-Mother or the illuminated dark mass or matter with a twinkling as the cause and not the effect.

I was going to ask the senior what everyone thought: "Which came first—the egg or the chicken?"

Both, as part of Universe-Mother or the total (everything existed before with or without the big bang, or infinite big bangs), which could be the past and present, because the past is the present left behind as history and the moment is time that creates, registering and living behind it as it rolls continually into the future.

Time does not stop; it is imperceptible to humans, even with their gadgets, because it grows and ages, transforms and turns the celestial's colossus at a speed calculated by our science of a femtosecond, or better yet in a *sextillionth* of a second, opening the frontiers of the future, always transforming and building, but always on the present time, real as the sunrise and sunset, proving earth is moving on this fabulous phenomenon, imaginary as we feel it, as gravity, as we notice the effect from the cause.

Before the big bang, after it, or without it, the Universe-Mother or the total always existed before the universe, because the universe or universes are part of it as elements, as the atoms of Einstein, as part of the composition of all matter, as the stockroom of construction from which to create. As you all notice, my simple explanations are for the general public to have some idea of the marvelousness of our existence and not for the "brilliant minds" in the world where there is no Creator. They, in stupid arrogance, rebel against the rules that come with our existence as part of it, and the ones that affirm there is no afterlife automatically condemn themselves to having a miserable life, besides the normal things as good or bad to all of us. Some people have bought my book *How to Suffer Happily* (2001) as a challenge and then began happily suffering, because it doesn't stop the suffering but helps us accept it as an inevitable road. Let's then travel doing our best to smooth it.

The blast of the big bang was followed by almost infinite fragmentations or explosions or a combination of both. The material concentrated on the egg, which had everything necessary as a total package from A to Z, including the atoms that compose our bodies and spirits. They were and are arranged from a supreme mind and not by themselves. An illiterate knows it, but not the "brilliant minds" who want to impress the minority lost in their ignorance. They believe they carry the aura of "Caesars."

The material was concentrated in the egg according to theories from a few in science. I have the feeling it came from an egg in our noses called the ovum as the microscopic sperm enters it. The process ended as a big bang when we all emerged from it as a baby (human being). Otherwise, the sperm wouldn't rush on the race for the ovum. Then it finds in it all necessary to face life outside the womb. In this combination, men and women, neither one can say who's more important, because if you have the fish but the water isn't available, then the fish will just die. The fish, the fisherman, and the pond owner would die of starvation.

If we are not born, there would be nothing, because we are the ones who make realities as intelligent beings, existing to appreciate what was created. "To be born or not to be born" is the mystery at the hand of the Creator. You and I are created and must be thankful for not being just a pile of atoms. After being created as an energy called a soul, then we go into a sperm and egg to begin our carnal odyssey, as eternal spirits.

Stephen Hawking should have spent more time helping medical science to bring relief to calamities, including his, and not seeking black holes from deep in his "brilliant mind," blasting the one who created him. The purpose of his ordeal of being disabled could have been his mission to put all of his "brilliant mind" to helping others right here and not seeking black holes and going after the infinitesimal, as it should be by those not in his physical condition. He could have a home telescope to pick on it for fun, as I did, while living in Miami, North Miami Beach, and later in Oakland Park, as time rolled on in my life.

His pride as a black hole man is not helping him, as far I know, but he should have given us the defense against this anti-God monster. If one gets off track and heads to appear now, he and his family would became black hole food, because there will be no God to defend him. This galaxy's devourer terrorizes angels and demons alike, including small children as they have nightmares. This is not a brilliant idea, unless he was competing with Walt Disney to entertain people.

Everything in the universe relates to education or evolution for the better, and if someone wants to go deeply into any subject, all he or she must do is jump into the subject or matter. At the age of sixty-four, loving airplanes like the average man, I finally stopped listening to the family and jumped into a flying school, getting all the knowledge, including how to pilot my own plane alone with God on the right seat (reserved sometimes for instructors or a guest).

As I had promised Carol, my deceased daughter (1964–1996), a few months before her death, I was going to quit my aviation hobby on the day I did my *one-thousandth landing*. I would give up my pilot's certificate. When my wife and grandson added up the pilot log book on the calculator, it had 1,042 mostly night landings. I had satisfied my dreams. I was the first pilot to give up his right to fly, because keeping a promise is honor and Carol is a spirit. I honor her as much as when she was among us, personally as a young, beautiful lady. I should get the Nobel Prize for surrendering my pilot certificate. I was either a hero or a fool, because I could have kept it on a pedestal at home. The temptation was so great, like an alcoholic staring at a full bottle of a fine whiskey and not succumbing to the vice.

The same thing happened and continues happening with all of us. One example is Hawking going to Oxford and Cambridge Universities to study cosmology, giving him more knowledge. As he aged, an illness crippled him. He then could dedicate all his time and mind to bringing us solutions for medicinal needs here and now and not billions of miles away.

Anyone can do and say it, but what counts are positive results that help humanity and not theories, because the great minds are the ones who have fewer theories and more results. Thomas Edison gave us the lightbulb. Otherwise, we would be still going to light the kerosene lamps.

We need brilliant minds finding solutions for energy that can turn the gears to move our machines, especially cars, as petroleum is going to disappear sooner than we think; better yet, we need them working in all the fields, including feeding the world and dealing with the warming of our only planet, and not staring at infinity as I did, astonished for a few hours and not acting as a professional, while cancer is killing millions every year. The list is endless.

We must look at infinity and wonder. Gazing as a hobby is great, but being obsessed with it and bringing fantasies and warnings of destruction beyond our ability to defend ourselves is just ridiculous and child's play.

We have been to the moon, and Mars will be next. It is evolution. It brings deep technology for a better life on earth, but we can't even live on either one, because our bodies wouldn't take it, and the fuel spent to get there is not reachable to most people. Let's stick to trying to save the planet. It is getting sick, and it could be lifeless even before our grandchildren get married. Our future after a few generations could be zero. We could become like the dinosaurs—extinct. There is no need for an asteroid, just a few atomic blasts when North Korea begins and the United States ends the fireworks.

Science did a good job creating the atomic bomb. It could create a calamity that could be the end of our planet at any moment, as could the deadly gases, which are still in large stock. Now, there are more black holes and a billion asteroids heading toward us. Is there any hole to jump in? If so, is it at the cemetery?

I am not worrying about myself, because at my age, the soul should be ready to go. There is no luggage allowed but good deeds, and this is what Hawking knows; this is what worries him more than his black holes, because his surviving the crippling stage was a miracle and his time isn't up yet. Aging is no way out. It is getting closer. I am packed with good deeds and absolute faith, and that's the spiritual balm.

Beyond our solar system, billions and billions of miles or light-years away, on the reflections of lenses, some nut is alerting us to black holes while telling us to forget about the untruths of a Creator. The marvels from A to Z were merely puffed from nowhere and restricted in the human mind. It is supplied with fuel, like any other irrational, but with a fine grass, and it should give us thoughts of better behavior. It is not in everyone's agenda, as many of us only dream of being in the fake spotlight at any cost.

The only way to reach those worlds is as a spirit, and it is not a theory. I based it on common sense, not religion or because someone told me. Hawking or no Hawking, I accept the challenge from anyone.

With the contamination of the planet or the sun getting too close to roasting us, we can never get to rockets, as we take trains at the station in the direction of the celestial globe, because the Great Designer did not put it in his agenda. It is the reason he created death, as our freedom to reach the galaxies.

The grandeur of the universe can be seen through our windows with telescopes. It gives us the grandiosities showing that it is the Creator's postcard, giving us hope as the only medicine we have in confronting such a beauty. Our intelligence is everything for us to feel our eternal life, beginning in the conception of the sperm and ovum. It is evidence here—not in the reflection of a distant corpus but at home under a microscope.

Without the Universe-Mother, we would not be anything, not even the *nothingness*, because it is a frustration of minds unhappy with the creation, having death as the end of the body and afraid of the dark. Whether someone is intellectual or not, he or she lifts his or her nose, looks at infinity, and says this or that as experts. I listen based on theories, and I take their word as Hollywood fantasy, not to be taken seriously, like the prediction of the weather, where the rain is coming, but suddenly, the sun shines.

In this hiding and silence, God is fragmented and camouflaged in all his creation as part of the Universe-Mother or in the dark mass. We see it with our bare eyes on a clear night, encrusted with stars. We can feel him in our thoughts, when we are alone and do wrong. We even looked in our bedroom or the middle of the desert or flew alone at thousands of feet up in nowhere. We feel like we are being watched as we all look around and up and down. We feel like we are not alone, as our conscience bothers us.

He can't appear in a multitude, or he would be seeing and listening. That's why he seized an eighty-year-old senior, author, and journalist, who is talented in many fields and worked hard to have a better life and while doing it, helped others. He has suffered through his heavy weight and moral cross. While this is going on, all the endeavors of life can offer spiritual tests and never blaspheme life or the Creator. He lifts his spiritual baggage, knowing education is the only thing that will survive the grave with his spirit.

But as the years went by, his learning was at most having a Creator. It is beyond question. Only a renegade would not agree, facing the creation that came from nowhere, but it is solid from a fountain, as it is the life of all the existence.

Said the senior, "To know is to seek, and to seek is to read." Researching in all the media, traveling and talking to people of all walks of life, in all fields from material to spiritual will allow you to know the terrain you are in, giving a basis to life. It is better to know you are not alone. Men not only need bread to live; otherwise, he would be like the irrationals, but he gazes at the dark matter and sees its infinity where light meets the imagination, because his is a share of the chip.

"All the glories on earth," continued the senior of the galaxies, "are spiritual, because material glory is short-lived. It dies with the carnal body, but in conjunction with the spiritual one, it interlaces with the moral. Together, it creates hope as a bridge. In this ecstasy, the soul as a spirit reaches beyond death, because we can't die twice."

2 Observation

Dear Readers, when you get to the end of the book, you will have all the logic from this old senior from the cosmos. The cosmos is now an elegant word. Carl Sagan with his marvelous programs put the universe in our homes, and Hawking jumped on his wagon, as he always does. He has time to do it better, as his monitor offers him all the necessary data. As far as I know, the cosmos is the frontier beyond the universe's border, as by now, Sagan, as a spirit, must be there. From there, there is easier access to Hawking's universes.

Will it be a ghost from heaven, a spiritual prophet, an angel, an unlighted spirit, the Christ, or God himself disguised as a young person? I bet the majority will go to heaven as angels; they are so numerous among us that when someone offers love and charity, he or she gets the nickname of being from heaven.

Angel statues are all over. From cemeteries to our first costumes in nursery school, we get our first wings with feathers, but I changed mine at age sixty-four for the metallic ones of an airplane. I guaranteed I would get closer to God, but I found out my wing tips had angels holding them for a sure way of returning to solid ground or I would not be writing this book now. On my recent trip to Lima (Peru), they had beautiful, cute clay Indian angels, and I bought a dozen to hang in my ground-floor apartment entrance. The next day, they all flew away to Peru. Next time, I will keep them in a cage with cameras and alarms.

I love you all.

William Moreira (Canno)

3 The Nail and God's Dog!

The farmer was sitting down on his spotless porch. Sitting alongside him was his big beautiful dog. The dog was desperately suffering in agony. A neighbor was passing, and he asked the farmer why the dog was suffering and what was the cause.

"Why, God, is your great, gorgeous dog in such pain?"

"He is sitting on a big nail!"

"Why doesn't he get up?"

"It is his free will, his choice. That is my number-one gift to my creation as my offspring, including accusing me of injustice and creating wrong in the universe, suffering, and death. You do know death is just material. What hurts is some of my children say they are fatherless, as I do not exist, because I am not around the corner giving favors. But, meanwhile, you all fight each other in all kinds of arenas like irrational animals. The wars begin between four walls and end on the front lines. Hell's doors have a backup of bloody souls."

"Dear Lord, why do you, as God, full of wisdom and generous, the creator of the Universe-Mother and the big bang, not help this poor crying dog?"

"As I said before, he doesn't get up because he is hardheaded and a stubborn and shrewd soul. He hates God, and only time and pain will make him come back to me. Satan and his family of demons are getting closer to my feet. Just wait!"

"But, my God, no one loves pain. Eventually, the dog will get up?"

"Yes, they all do sooner or later. Some last more than a half century as a show of ignorance against themselves, even believing they are above me in intelligence, but they can only blaspheme not create, while sitting on a divine nail of reprehension, as fools in their free will. Pain is a reward for bad behavior."

This parable came to my mind many years ago. I noticed on TV the stress of suffering and almost an absolute paranoia from the famous scientist with his cosmology and quantum theories Stephen Hawking, now in his seventies and immobilized in his infirmity. He was completely paralyzed, and survival was considered impossible. If he had looked for a miracle, he could have had a surprise. If we do not put a dollar on the lottery number, we will never win. God is in no hurry to have Hawking come back to him, because he is going to spread rumors in heaven. He is the street cleaner, because of his 250 IQ qualification by sinners on earth.

Considered by some as a brilliant mind, as a master of theories on the universe, he followed the studies of Edwin Powell Hubble, a giant in perfecting the telescope. He was the genius who went beyond nebulae to the galaxies. Around 1929, he affirmed that the universe was expanding and Hawking jumped on the same pedestal. The lines of many stayed on the same wagon as Hawking. He became a martyr of his immobility in his wheelchair and his sand pedestal. When he used his followers, atheists stepped in the quicksand of a creation without a Creator, affirming in *The Grand Designer*, a lifeless universe where everything ends seven feet underground in a muddy dark hole called the grave. Even philosophy died in a black hole—not millions of light-years distant but in his own backyard.

I did not complete a doctorate in astronomy, much less cosmology, because I had other priorities and did not just look in one direction. I got enough knowledge or know-how to dream and most of all to enjoy life with my children as our right, while being in business among people, creating jobs and spending to keep the economy rolling. I did not sit around imagining a world inside my walls and having my ideas about the universe. I did not stare at a monitor but the real thing, alone at fifteen thousand feet even when it was ice cold and transparent. I was close enough to feel the twinkling of the dark matter, because I had "contact," something few are able to obtain.

The world is too great and beautiful to be concentrated in one degree. As a pilot, I enjoyed the 360 degrees of freedom in a window at fifteen thousand feet above ground. I liked looking out in the wee hours at the total or the Universe-Mother as real. We would have to have been created. Looking at only a few thousand miles of the real picture, at the curvature of earth in living color, beats having one's nose in the plastic images on a computer screen.

In the skies above Florida, New York, New Jersey, Pennsylvania, Delaware, the New York sound and the islands, and California, passing over the Rockies to Las Vegas at night, I could feel I was in my own private celestial world appreciating what few dare to do—not as a daredevil, but as a humble human being feeling the creation not sitting in front of the media giving false logic as logic without getting my feet wet.

I was not satisfied looking at pictures of the stars on paper or TV while listening to private pilots talk about how they enjoyed looking at the real thing. Able to do more than just talk, they can fly themselves for their own gratification. I did it proudly, but it is not easy. It is dangerous. Our lives depend on a machine not to break down. If it does, death is almost guaranteed, but the reward makes us feel the Creator. It hurts when someone sits on the ground just blaspheming. It is like the ones throwing rocks at a caged lion. This lion did not roar at me but gently said, "Son, you are welcome; the angels are holding your wings. To be sure, you made a perfect landing on this windy night on the ground."

Hawking's theory of nothingness without a Creator is his opinion. It is his right. It is also my right to have an opinion. Everyone has the right to express what he or she believes, as affirmed by Thomas Paine. He was a genius who used his powerful political logic, not theories but realities, for our rights. Hawking kills hope, and now that philosophy is creating negative feelings in human beings, who are already carrying a heavy cross. We are not in heaven where there are no wheelchairs or crippling illnesses. I know for sure there is no effect without a cause. It never fails.

Thomas Paine was an Englishman (1737–1809); he was one of the most intellectual men. He made history with his pen, not writing novels or scientific theories, bringing black holes to terrorize people but light to everyone's soul for a better society. He wrote about equal rights and justice, which cost him his life at the age of seventy-two when he wrote *The Age of Reason*. Now, his name remains a part of history, while the "brilliant minds" are forgotten a day after their funeral.

We do not choose our cross; it is built to fit everyone's needs, or suffering wouldn't be logical. According to Jesus's parables, he said, "If you want to go to heaven, pick up your cross and follow me, because mine is light!"

That means if someone offers you the key to heaven because he or she has connections, it's not true, because nothing comes free. We have to work for it to deserve it. It is more than logical, because when they offer more than money can buy, it means something is wrong.

We do have intelligence above irrational animals that usually have feathers, fur, and defense systems for the protection of their bodies, while only eating for survival. They are born, grow, eat, and have sex for procreation, and their grave is in our pots and grills.

We can chain the body but never the soul, because we have freedom of our thoughts, which proves we are intelligent. Intelligence is logic and common sense. It does make us understand there is a Supreme Intelligence. The mediocre is distinguished as a "brilliant mind," creating theories as realities about things not in their backyards or even on earth but at distances that we can only reach in our dreams or as spirits.

Meanwhile, the equations of realities are right at the tips of our noses. They continue as mysteries to be resolved by religion and science. In the meantime, the microorganisms and parasites are winning the battle in the form of sickness without solutions. To make things worse, the waters are becoming polluted and the atomic arsenal is a blade, just like the guillotine hanging above our necks, ready to cremate our earth, our home, our only home.

We can defy our Creator with our free will. Then come the rules of profanity, which is a maximum immorality. It is a price to pay. It is not an eternal inferno, but a temporary reprehension that could be a full life in our carnal bodies. The only alleviation is to begin a new life in the spiritual. That's all I can say. Your imagination will finish what I meant to reach into your common sense.

Writing a word in a few minutes by speaking like a robot into a computer and dreaming of multiple universes and celestial bodies would be hell. It is hell where there is no intelligence behind all this imaginable existence. It is to have a life encircled in a casing like the butterfly. When they are all free in the wind but you are a prisoner in it, unable to reach out, there is no way out and your reaction is to keep denying there is a Creator. You know that is impossible, but why not ask him what you have done to deserve it?"

I am thankful to our Creator the moment I open my eyes, get up, and am able to walk, and I accept everything else, painful or pleasant, with understanding and humility, even the moment when I received the call from the New York police on February 5, 1996, saying my daughter Carol had been run over by a twenty-two-wheeler and died instantly.

I continue up to my eighties, seeing most of my family and friends die, and I always said and say there is a reason for it. I include the good, like having good health, the miracles in being blessed, my trips to dozens of countries, enjoying different ambiences and savoring food, and my dream of flying at the age of sixty-four. I thought it was over, and it had become reality, in just the nick of time.

It is like being lost in a desert and a gorgeous city pops up in your way. When we have an understanding, we are the created and not the Creator. We see all the goodness offered to us and feel the freedom of free will. We are human beings and must put it all together. We feel like we are looking at beauty, such as in the smile and love of a child, a gourmet dinner served pompously at the Rainbow Room on the top floor in New York, Italy Terrace in Sao Paulo, and Galata Tower in Istanbul. (I did a portrait of a beautiful young lady, over the linen table with my *Parker 51*, and I didn't have to pay the bill.) This inspiration was a half bottle of good wine as I stared at the beautiful young lady and the Creator's sky. It was a privilege to know we are not born to an existence full of negativism, because it is more than that.

On the last three cruises, where my last two books were born, I appreciated from my little table in the buffet area on the top deck the immensity of the ocean and the golden solar rays striking the small waves and balancing the vessel, making my pen run more easily over the paper. The words exploded onto it. I asked the Creator that the words on realities would help those lost as offending hearts.

I felt peace in my conscience when dozens of people used to come to my table and look at my books, saying they were blessed in having met someone as spiritual as an angel in a world where the devilish are taking over and few try to stop it.

During the ten days sailing toward Buenos Aires and back to Rio de Janeiro, I was noticed by the 2,650 souls aboard, and they all wanted to know what the book was about. I just gave them some pages to look at and my life on board became like a paradise. Now, on February 24, 2013, I have been sailing for seven days on the Brazilian coast on the *Costa Favolosa*, but I will take with me my *Parker 51* and the thick notebook just as a precaution.

Those passengers came from all walks of life, and they questioned me about the big bang. Was it the beginning of the universe or universes? To my surprise, a few mentioned the total as the existent dark matter or as eternal as the former universe—or even better yet, the birthplace of the universe. That was when I had the idea of the Universe-Mother or Mother-Universe.

I feel as if we are in an endless aquarium where everything is in it, including the chicken of the egg of the big bang, and a huge octopus is eating all the fish up to the blue whales. It is called a black hole as a master creation, where his creator is free in his imagination, terrorizing—I mean theorizing—a universe with a vacuum cleaner eating itself. Hollywood took advantage of it, making two movies about black holes, as people run desperately from the theater, but someone was enjoying it in his world as a "brilliant mind."

When I mentioned Hawking, they all said he is the creator of a monster that one day will eat the universe, the whole universe, with its voracious craving so that not even light will be spared. Some commented that Hawking needs psychological help, due to his physical sufferings. Demoniac thoughts are now part of his fantasies.

To write about something, first, we must get all the facts that are necessary; we must research, analyze, and seek education as much we can to do a fair writing, as justified as it can get. Realities many times hurt, but they heal as medicine. The bitter-tasting ones are the best.

The whole existence without us humans to feel and appreciate it would be just a pile of atoms. Whatever others' compositions, pulsing and exploding, creating a senseless mass of combinations eternally with or without light, it would not make any difference because there is *no one to feel it and enjoy it*. That is when we came to the stage.

This was the Creator's frustration when he finished the marvelous creation, which included all the irrational animals, millions of which have now vanished. He gave them limited intelligence and instincts to survive. There were no family ties or any sentiments to a level to look up, offer thanks for the star-encrusted dark canvas, and in admiration say, "Thank you, Allah! You are great!" and then fall on their knees, hitting the soft sand of the desert, putting their faces on the carpet at confronting such celestial beauty.

He felt lonely for the first time. The irrationals didn't even look at the gorgeous sunset or admire the shooting stars; they remained with their noses in the grass or their fangs in flesh, fascinated by their meal. Then, the Great Designer went to the calculation table and in a second put on it his own image. He called us rational and gave us top intelligence to become his family, and then the big bangs began popping all over the dark matter. Everything else came as a package, as saints and demons to keep us alert and alive.

God created the universe and his family to appreciate his marvelousness. The biggest example is in the *réveillon* or the turning of the old year to the new one on Copacabana Beach in Rio de Janeiro, right under my windows. About five million people stand shoulder to shoulder as brothers and sisters, including beggars, millionaires, and all the millions of tourists from all over the planet. In the background are a dozen cruise ships, while Christ at 2,200 feet is blessing the sinners, blasphemers, and the ones lost in their ignorance because the Creator sent him to us. He wants to have all of us as children in evolution, in this fabulous concentration of souls and not even a wallet is picked that night.

At midnight, at the first burst of colorful fireworks, covering the whole area of the four miles of the curvaceous sandy beach, the clamor of the multitude is heard from hell to heaven: "God, we love you! Happy New Year!" (*Deus te amo, Feliz ano novo.*)

I sat all day long at my little table on the top deck with one eye on the ocean and the other in the paper. I was interrupted by passing passengers, and I loved it, because without human contact, there would be no education. The diversity of questions keeps life colorful, and my pen never minds it.

I just found out. It takes a lifetime of extraordinary effort to understand how important we are as human beings, related to the infinity of creation.

The average person says that we are too small to be counted, especially religious leaders, making us feel we are nothing in the Creator's eyes.

I hope those with these negative thoughts, which they believe are humility, read this book, because we are very special and important to the Creator; otherwise, he would not have created us.

He gave us free will because he needed us as his children to give him headaches. A house alive with intelligent children gives us happiness, and time flies as the day feels like a few hours.

Sometimes, as I am writing, the pen just touches the paper and the words explode into it, giving me time to fill the pages almost like the printer in my computer. This blessing, if I can call it a blessing, gives me the opportunity to keep writing the thoughts, as the mind jumps or moves from one thought to another.

Nothing is more boring than a book staying on the same subject, not changing its color, like music with few notes or a large dish of a tasteless soup.

I traveled to a few dozen countries, and it is fascinating and well worth the money spent. I like to walk in the streets without a destination, observing the diversity of cultures and tasting their culinary offerings, smiling at the coffeehouse and having conversations, not seated on a "nail" blaspheming a Creator as the guilty one in our misfortunes. As said a wiser man, "Before you accuse anyone, look at your tail first."

I am proud of making a great poster or designing for the feminine foot a shoe or sandal not imaginable by others, but when I look at the Creator's postcard, I humbly bow my head and say to him how great a designer he is and thank him for giving me the chance to scratch my own. (You can see a few I did on the cruiser on the last pages, including a beautiful, strange woman's face staring at you. It was my last painting.)

One young crew member from the cruiser staff looked at me and said I looked at everyone like I had had a good life, because even when I wasn't smiling, my looks affirmed that life is beautiful and there is nothing to worry about. I looked at him, amazed by the extravagant statement. I felt happy, as maybe he was sad for me in my age. He was kindhearted, as a good soul, and tossed me colorful hope, saying that to live is great. After I took a deep breath, I responded, "Son, life is beautiful in any circumstance, because we are God's children. We must look at it while on earth intelligently, and even if we are going through bad times, won't demonstrate it to others, because you could ruin their lives. It only takes one word to do it. Tears are not forever, but happiness can be forever."

The dinner was exceptional, and afterward, I went up to my top-deck table, and as I took the pen, I noticed over my manuscript a little dish of biscuits, a glass of cold milk, and a small note saying: "Thank you, Mr. William. We all love you."

I closed my eyes and felt the warm tears rolling down my cheeks. They were tears of thankfulness for our marvelous universe, for how important we are as part of it, as intelligent beings, remembering good times as they flashed in front of my eyes. I felt that what we give, we receive back, as a reward for being kind to others, and this kindness begins with the right meanings of words. The power of a word can be oral or written, and it can heal or destroy. Those responsible for it will feel the cause and effect as divine justice.

The past is impossible to forget; good or bad, it is in our spirit memory bank and follows us through time into the future. It stays in a file, and we can reach it when we feel like or circumstances pop up. The computer originated from it, but ours is perfect because the energy comes from our eternal spirit. That's why I love to write, because my past flows as I put words to paper. Amazingly, even dates are correct. The sounds of voices from family, friends, and strangers come back, as a merciful, precious display. Says the wise man: "To remember is to live!"

4 *Costa Concordia* to the *Grand Holiday of Ibero*

December 13, 2009, I was aboard the ill-fated *Concordia*. I began and finished *GOD! The Realities of the Creator* (260 pages). It was born in seven nights between hell and heaven, and thirteen months after its publication, on January 13, 2012, *Concordia* sank because the captain embraced the alcohol called Satan. This tragedy happened exactly one century after the sinking of the *Titanic*. In 1912, the designer of the *Titanic* said not even God could sink the cruise ship, and it went down to the abyss of hell with him on the first night. He paid dearly for his blasphemy but dragged many souls with him. You should feel better; the effect is without a cause because God isn't unjust.

Stephen William Hawking is vainglorious as a "brilliant mind," as a master, know-it-all, especially in cosmology, and as always, his assertions are based in theories, beginning with quantum. He sits now on a big sand pedestal on his discovery of black holes that go around the universe destroying eternally even light, making God an unsuccessful Creator.

Everyone who has read about his black holes feels powerless in knowing that up there is a monstrous, voracious galaxy-eater. They run to their church for protection against an evil loss or coming to some ill-fated end. He will spend his entire life mummified in a chair. He should, as a Catholic priest, begin a chain of prayers to help his soul. They are a Good *Samaritan's* thoughts.

Worse than the black holes is his affirmation up to now that there is no Creator, just a void or null. That makes me question the ones qualifying him as a brilliant mind. Every cause has an effect, affirms science as based in reason, logic, and common sense. His rebellion, I bet, is based in his physical condition of being alive for over a half century as a mummy. He puts all the blame on the Creator as the cause of it. If he takes his eyes off the monitor and looks at his tail, he will find the answers. He is quoted as being a smart monkey and so he knows about the universe. It isn't a joke. Just search "Hawking's quotes" on the Internet and have fun.

In the United States and England, as well as in many other countries, for defamation or slander as well blasphemy if it is proved by an eyewitness, written, or recorded, the guilty one is immediately arrested, and the consequences are harsh, including years in prison and a monetary reward for the victimized. Those negative accusations can destroy a life or business, and many times, they cause crimes and suicides. Those with diabolic minds or tongues are worse than the wipe of hell, because they take hope and dignity from souls.

I said that maybe this is going to be my last book because the travel insurance company, covering health, life, and luggage, sent my policy with everything in red, but I expected my luggage in black. It was the only insurance I could get, because I was over seventy-five. I accepted it, because many times in life, a little is better than nothing.

Once again, after two years and also in December on the same square in Rio de Janeiro, with a five-hundred-page notebook, several ballpoint pens, and my *Parker 51*, I confronted another white swan from the galaxies, the *Grand Holiday from Ibero* but with *Costa*'s magic inspiration and culinary delights, which began with *Concordia*, the ghost. It is not a coincidence that it sank exactly one century after the *Titanic*. It did not hit an iceberg but a rock. The results were the same. No one needs to worry though; the captain of *Ibero* was not a blasphemer or an alcoholic.

Aboard, I felt affected by the mystic atmosphere of being in it, sensing the movements of floating where gravity and antigravity fight the heaviness of the vessel doing the impossible making it possible and avoiding the sinking process.

In this process of nature, it is like being in three dimensions and not feeling the solid ground. Worse or better yet, when I piloted, we were thousands of feet above sea level at zero altitude, and the air wasn't as solid as the water. The relativity of senses neutralizes the soul on the spirituality of freedom. Well, I do my best with words, written or verbal, because there is no way we can express exactly what we have in mind. How can we describe a cruise ship to someone like an Australian aborigine who has never seen one? How would you even describe the ocean?

I was among the 2,650 souls, hearing their opinions of happiness and sorrows, because as a senior with white hair and a beard, writing books about realities, I felt like I was their father or grandfather with enough celestial wisdom to agree and give them some right answers. Sometimes, I felt like a "lamentation wall," listening to their whispers of sins and hoping for a backup in their actions. They were expecting answers to the mysteries for comfort in a tough world where solutions are not around the corner.

The direct contact was based on eye-to-eye interaction; it makes me a better writer and journalist. We become more logical as humans when we feel our existence as souls in the inescapable world of spirits. It's logical and better be logical. Most of the people I knew in my eighties are all gone, and also the population increased from 1,933 up 2,013 from presidents, kings, and queens, movie stars, religious leaders, and so on. They all fell from their sand pedestals, right into their graves, and the only baggage allowed to float away from it with the spirit is the good deeds of their conscience.

Before I got to my senior years, people's complaints used to irritate me. I would tell them to go to any religious institutions or to Israel, because the Lamentation Wall is rock. It could take their problems without being irritated. As the years rolled by and my own problems gave me endless nights of worries, I began feeling compassion and wanted to help others. I was sorry for the way I had behaved. I notice that as the majority of the population gets older, the ripeness that comes from pain and problems isn't always enough to soften their hearts, while the line at the "Lamentation Wall" in Jerusalem goes up to the airport.

I notice that bad news is free and burns like a fire in the dry fields of hell. There is no way it can be put off. That's why man's law is based in God's moral law. The blasphemers are in the spotlight, and the payment begins here and finishes on graduation day.

On the top deck of a cruise ship, I felt I was between hell and heaven, confronting the ocean at night and surrounded by the illumined dark matter of the Universe-Mother or the total (a rarified, neutral, unknowing twinkling mass to us, as everything came from it, like we are in a jelly-aquarium and everything originates from this gel. To live, we must be on earth as part of it).

We can see from the cruise ship, far from city lights, all the dark infinity. It is so clear that we can feel its deepness, as it has no beginning or ending. People like to say it is encrusted with stars, but really the celestial bodies just float in it, as we do in the water while scuba diving. That was my favorite sport when I used to fly to Key West. Coming back in the wee hours, I liked to feel the marvelousness of the view, which is not available at any computer monitor.

The theory of Hawking's black holes, my young granddaughter said, could be because his only vision is the black background of the monitor; after years, his retinas started to swirl the darkness, and it became like holes. When he had access to a telescope, he saw the swirls of colorful masses as galaxies or whatever.

My granddaughter loves science and has top grades in mathematics. When she says anything on those matters, it is because there is a cause behind it. Life is better as realities and not just dreams of hole theories. She's not sitting emotionless in a wheelchair but running and jumping as we all should.

Since I was a child, traveling between cities, I have just opened my mind, amazed by the differing darkness of the infinite different positions. The mass isn't really black or dark; it is a composition of a kind spectrum from a reflected glass, but then came the word of rarefied, esoteric, or holistic theory. I sincerely believe there is no word or combination of words to describe what the eyes can see.

When friends or relatives, including my wife, called me irresponsible for flying at night, I insisted they should go only a few minutes to see my craziness at night. A few minutes is better than a few hours, and many did go up. They were in no hurry to come down when staring at the infinite blackness, as they loved the spectrum of infinity. My wife became my companion on a trip to Atlantic City for a midnight coffee and hot pastrami sandwich at Caesar's Palace. The small airport was just in the backyard. The beauty of the night sky concealed any fears. It was too bad Hawking couldn't enjoy my little adventures, because I know he would have found the Creator. The design is too much to ignore, even with his black holes.

The up and down motion of the bow and the sinking of the stern made it feel like we were asking the Creator for clemency. It gave us the feeling of fragility in a beautiful life in any circumstance. He gave it to us. If it wasn't for us to appreciate the Creator, he would have had a hollow existence with all his supremacy, because there would be no one to "break his chops," as they say in English, especially now with a family of billions, as he miscalculated the power of sex.

More than ever, the great telescopes have opened our windows and dreams to heaven, bringing to our eyes God's backyard. What is not reflected in the retina is hard to believe. The old saying, "To see is to believe," indicates that being an eyewitness puts that extra nail in the coffin.

The atheists or unbelievers (I don't believe in atheism; it's for me a show of scarcity of knowledge about the phenomenon of death) are like many friends of mine. That's why the incredulous are the ones who own their own private telescopes. In the wee hours, after nightmares, they run to it, desperately seeking the Almighty. They blindly see his postcard. The same goes for the speeding driver doing 120 miles per hour before seeing the sign there is a curve and to slow to thirty; they smile and then wake up in hell, but it is too late.

The whispering of the wind and the bow cutting the small waves make me feel how small and big we are as parts of the universe, as being originated by the Universe-Mother. Intelligently, we see and sense the grandeur of existing eternally, because the religions say we are souls and science confirms us as spirits without frontiers. What more do we want?

Authors proudly confirm having been hiding away from sounds and contacts to concentrate the mind and direct the pen to the paper. I continue being amazed at differences. If I do the same, no words will flow because of the absence of pulsing life, but being surrounded by classical music, the ocean, or beautiful scenery makes my day and is my inspiration. Beauty is involved in teaching the mind that life is great and everything was intelligently organized for us. Here and then, life is as timeless as time itself.

On the ship in midocean on a crisp, cold night, between Brazil and Italy, I could see the paper without the lights, because the light from the moon comes from way above. It is powerful and eternal, and the words just come bursting out from my old faithful server, small, sleek, and powerful, because it has not failed me since 1952. It was my first investment, lasting a lifetime, my *Parker 51*, with true ink. Unmercifully, they stopped manufacturing it, but I have it as an icon of how a pen can be. In the right hands, it can build or destroy lives and even nations.

Life on earth is so interesting because not only could the next day bring big changes, positive or negative, but even the next minute, and the same goes for our staying here in our bodies. Just two years ago, I embarked upon the *Concordia*, majestic as the white swan of the galaxies with its thirteen decks, representing man's conquering nature, carrying five thousand souls in great luxury. We belonged to another great dimension of the universe, where the 112,000 tons floated like a rice shell as solid as a rock pedestal. Then it puffed out, like it was being navigated on quicksand, just like a billion-dollar aircraft that is said to be the state of the art but falls from heaven like a mortally gunshot duck.

The interesting issue is whether it is our fault or a destiny prearranged prior to our tour on earth as human beings? I will let you e-mail me the answer! And I will answer you back. (The e-mail address is in the final pages.)

Majestically, *Concordia* was ignored, defying God and Satan, forgetting its fragility in being the created and not the Creator. The "brilliant minds" are only small minds; they are a grain of sand on the beaches of the universe compared to the Supreme Intelligence.

The irresponsibility of egos only brings the abysmal suffering as a reprehension to moral values, and it includes all of us, with or without status on any level. We were created equal, but we have the free will of choice to learn and upgrade our values. It includes morality, and the list of moral issues includes all the necessities for our spiritual journey. It is not a religious issue but common sense based in reason.

The word *anarchy* is not from heaven but created in hell by us in our free will. It is not accepted by the Creator, and pain at all levels will reign until we awaken from negative behaviors.

God is not listening to cries of sorrow, hollow words in prayers, donations in billions to the religious institutions, or any charity groups, or screams of pain because it is all material. Miracles will only happen when we look at a beggar lying down on the sidewalk, filthy and starving, just waiting for death, and we pass by and toss him a coin instead of looking up and stepping away. Or when we see a macho man beating up his children or wife, miracles will happen when we get involved, not when we say it's not our problem.

Our misfortunes are here to stay until we see and feel we are just one to our Creator. As I have said and will say again, I am not a religious leader or a politician, or I would be expressing my opinion, as realistic as it is.

It took me eighty years of a long life on earth to learn to go up to a solid pedestal, being able only after walking and flying in the 360 degrees of rocky, painful roads before I could give advice to others so they could avoid the ways of hard learning and find out we are all one family, not just your brothers, sisters, and relatives, but also the neighbor whom you never invited to your backyard, the one who fell on the street and you did not even call for help, and the one you ignored when you heard something wrong was going to happen and just sped away so as not to get involved. Meanwhile, our world is the hell Dante wrote of in his plays. He showed the two faces of our masks.

The *Titanic* was deified, and the Creator sank it in its inglorious destiny, and the *Concordia* one century later on January 13, 2012, followed it into the deep, because someone irresponsible with his free will had not learned yet the "why of life," and the pains in life didn't wake him up to realities. Then came the lessons when the moment was ready.

The captain of *Concordia* had the right conditions for the event to occur, as he was not alone in the corner bar: *alcohol and infidelity*. The chain of events also includes passengers, because it wasn't a cruise for saints or the Creator. It would just be a monster, like Hawking's black holes, going around gulping galaxies or whatever is in its path.

Now, aboard the *Grand Holiday* for ten nights, I was full of hope, because hope is the only thing that will never abandon us here or there. During my cruise, besides writing, I would have been a fool if I hadn't made use of its offerings of life's enjoyments, like good food presented with class at the gorgeous and enormous dining rooms. The ones who worked hard to save to cruise will feel their money's worth for the first time for sure.

I can't forget *Concordia*. It was an ostentatiously giant white swan from the cosmos, nestled at Rio's harbor defying God's gravity with its 112,000 tons of steel, waiting magnificently for the thousands of souls guaranteed enjoyment like a creator, just as the "brilliant minds" go around judging God's universe as immoral as they can be.

They all are resting in self-images as egos, unfunded in inglorious theories, taking humanity to an abyss of sufferings. They became their own victims and remained so until the edicts of values and moral values. That includes everyone in society. They all now come to the mind, because we are now in the billions, but we all are accounted for. The Creator does not tolerate anarchy; otherwise, the universe would not have been created.

When boarding this cruise, I told some friends that perhaps this was to be my last book, because when we reach the glorious eighties, the weight of the body gets heavier while the soul gets lighter. The bridge that goes beyond the limits of the flesh appears, blinking, on the horizon as an indication time for when the trip is approaching.

The interesting thing is that the younger ones not yet in this level of time don't understand the aging process, as it rolls continually in the present time, never stopping, moving so slowly on as in the division called Planck's time. It is almost uncountable on the smallest division of the second, but it is a mercy from heaven. We have no time to feel its progress, and we all accept it in still being young. No one ever broke a mirror because of being old, even saying: *"We are like wine; as it ages, it gets better. It becomes alive with a sense of taste."*

Last week, I met a ninety-two-year-old lady as she was correcting her lipstick on her cute little mouth with a pocket mirror at the coffeehouse. She remarked, "I believe I am getting old, because we don't feel the passing of time." Truly, I know this great divine phenomenon, because it comes with experience.

When we hear the call to go, we are not allowed to take anything material at all—but our *Parker 51*, because they never run out of ink or leak. The good deeds and all the things done in the name of love are spiritual and are allowed because they go with our conscience. Naturally, our spirit will never be left behind.

I was in the long line on a hot summer day in Rio waiting to board when the speaker announced at the entrance of the empty bridge: "Attention, passengers, welcome to a paradise called *Grand Holiday of Ibero.* We are beginning boarding, but first, let the nice-looking senior, with the white hair, beard, and mustache come in. Please give him a hand, because the plank oscillates a little."

I felt for my laptop and boarding luggage that carried my intellectual arsenal, such as pens, papers, scientific magazines, designs, and so on. They all must go along with me and were taken up by the young crew. I felt like a spoiled little child surrounded by grandchildren. For assurance, I stepped in with my right foot as I waved in gratitude.

I felt the coolness of the ambience; it was like entering another world of beauty. The change of balance in not being on the ground made me feel as though I was big and heavy. I mean nothing to gravity, as it's facing the water, resisting the weight like a little duck. The battle goes on as both defy creation, balancing its power. We are the winners.

My number-one intention for this trip, which I had already had in mind for several months, was writing this book. I was trying as hard as I could, using a lifetime of experiences, both good and bad, with all my intellectual resources, which cost me thousands of hours of pleasures and sleeping on this tough effort; if I can be definite in this manner, I consider myself to be the winner as a lone star, because whatever I know is mine to keep and to help others with here. When crossing the road beyond the lights, the ship gave me the feeling we were in an endless aquarium, like a little fish in the ocean without borders, without a beginning or an ending.

I woke up from my dreams when a pretty uniformed young lady grabbed my arm and offered a short tour of the vessel to be sure I would have no difficulty in those labyrinths. It was full of lights and signs. I could never say no to anyone trying to help me, and this time, I couldn't do it to a beautiful young lady. The stage was ready, and now I felt the difficulty of not putting an avalanche of words on paper. I would be helping not one but millions. It was a dream of mine to change this great planet, on which I have already spent what is considered a long time. I have my heart settled so I can rest my head with some understanding of "why life is like this" as souls.

As children, we dream of Santa Claus at Christmas, listen to fairy tales, hear about little angels, and read all the fun and innocent stories, taking us away from the uncertainties of an adult life until we grow up. But this time ends when we go from being a child to an adult; it is unmerciful and has no forgiveness for those creating uncertainty. Not everyone has the capacity beyond childhood. It doesn't matter if those disguised as saints are in religion, science, or politics, or have achieved success as singers, actors, or leaders, snobbish icons in a society that is enclosed like a perfect butterfly to be born. As Hawking unmercifully tries to take away our faith or dreams of an afterlife, as he says there is no God, he is the one not enjoying life at all, as the effect of a cause.

As the elevator went up, I closed my eyes, imagining I was on Concordia. Upon opening my eyes, all the ghosts faded away as they usually all do, but not forever, because they are in our consciences and in our past. Anything that exists must be real; otherwise, it would just be a theory, part of the world of make-believe, giving no solutions, only useless fantasies.

What is happening to me will happen to everyone, because we are all equal. I can hear the voices of people from my past anytime I wish, and naturally, it is all in our minds, including the smell of perfumes, even the taste of certain food, and all the positivism and negativism from infancy to the present moment, because it is saved in the computer of our minds. We can stop in the good moments and smile, cutting down the stress, because we are in the present moment. It is wherever we are. The future is ahead of us, as we are in the present. I believe it is not too hard to understand. Anyone can draw a graphic to see it easily in a picture, like people do for the family tree.

Aboard, as a senior without a companion, I noticed the crew gave me exceptional attention. When someone grasped my arm, I surrendered on the spot, as the floor had become a little unbalanced, reminding me of my Carol (thirty-three years old when she departed in 1996), because she was exceptional with seniors when they came to the busy coffee shop at the Marriott Marquis in New York City. I was much younger, sixty-four, and she was already saying that her care of me was going to be total as a senior; she was an angel without wings, but now she's around using the young people.

No one has to believe me, as I don't have to believe anyone, but a few times, when things got out of hand, I felt her presence in many ways, calming me down. If it isn't true, my beliefs are still working for my benefit.

After an hour aboard, I already knew I was in the right place. I told the young crew lady about my intention of writing another book based on realities and observations of science, spirituality, and materialism bringing hope where there is no hope, making life a hell without solutions.

I can also draw perfect curves, and recently, I discovered I was able to design great feminine shoes and sandals. Women just love it. I realize that all the times I had four pair of shoes (black and coffee brown), while my wife had hundreds and was always buying more. It was an obsession, and I was going to design a few for the book. She took my business card and said she was waiting for my e-mail telling her when the book was on sale on the websites. The crew passed time on their long nine months aboard reading in English (see the last pages for several of those designs, and you can take them to a shoemaker).

I never say good-bye to anyone, like it is forever. I say, "I'll see you later, or so long," because it gives a happy feeling, especially because one day, we will meet again, covering a sensation of eternity. Everything in the universe is round, and there is no place to hide.

This roundness begins with atoms and ends in the colossal corps of infinity, as not even a square has been found yet. Even for Stephen Hawking, his only distraction is the universe and its secrets. Even his black holes are round, because swirling and curves are easier to do, follow, and design. I am good in designing (see the last pages).

After being welcomed with so much affection—and not only as a job trained staff, because we can feel the sincerity of people's souls—I decided to change my writing from nights to daytime among people. My open arms welcomed anyone for a short chat or a few questions, even while I had the pen on paper. I talked about politics and deep religion, so no one would be negatively excluded. I just concentrated on my sentences, avoiding negativism for everyone's benefit.

I can concentrate with pen on paper, even among hundreds of my new cruise companions, and in the dusty, dark caves illuminated only by cheap candles, in monasteries listing to endless murmurings of cantos, or deep in the forest in a shack while surrounded by mosquitoes, where my mind should be as senseless as those thoughts.

When we have a real inspiration, it comes with the contact between soul and spirit. It will flow anywhere, anytime like a broken pipe. Thomas Paine, at the age of seventy-two, wrote *The Age of Reason*, his last book, as he watched daily from his first-floor prison window (the Bastille) as heads rolled off from the guillotine blade. It was a warning he could be next, but it gave him more inspiration to begin and finish the book. His other great writings were on tiny scraps of paper. It was a best seller and not like the ones with their names on the cover, as the end is on the first page.

I exchanged the glistening of the galaxies for the sun's rays running at 186,000 miles per second to reach my papers, as the Creator's blessing on all of us. Night or day, we are in paradise as soul or spirits, and it is more than we can request and plead for.

It's 5:45 a.m., and the bow is very mild. It is up and down, waiting for a full day of positive energy, as the sun is busting up the rays illuminating everyone. It is its eternal job. The bow and the oscillations on the sides defy gravity against the antigravity. It is the relativity of the superficialities related to the vessel on the water and its weight, and it was not men's creation, but a handout by our Creator to our science. Once again, science knows this calculation works but not how it happens.

It's a continuing Ping-Pong game in our life, but the majestic creation always wins the disputes. As I was saying, we have the right to enjoy life in our glory while we can, because we were created intelligent. Otherwise, we would be eating grass and raw meat and not enjoying gastronomic cuisine and worrying about who looks better and worse, like foolish and inglorious beings. We accept others arrogantly, as brilliant minds or geniuses, while the average person fights with all his capacity to finish school with good enough grades to get a job as a janitor.

It irritates me when I see stickers on a car, written in color:

"My Son Is an Honor Student in School."

"We Are Proud Parents because Our Daughter Is Top in Her Class."

"Our Teen Boy Is Top in His College."

"God Blessed Us, as Our Children Get Top Grades in School."

But worse yet is Hawking's wheelchair:

"Here Sits a Brilliant Mind, a Genius above Anyone's Intelligence as Nature Did Its Best. He Found Out There Is No Creator. His Body Wasn't Created by a God, but Is a Freak of Nature, Doing a Miscalculated Job."

The whisperings of horror in our world originated from an egg without a background or, better yet, the first chicken-less one brings anarchy to kindergarteners and first-graders. It's like we were born spermless. It is like being fatherless, making a chaos of the Creation. It confirms that bad news is free and travels fast, because an egg that originates not from a nest in a chicken coop but from nowhere is so scary that it makes the black holes real galaxy eaters, Satanic beasts, as real as they can be, making people wonder what is inside the soul of this creation.

Interestingly, in Brazil, when the digital X-rays show clearly the fetus skull has no brain or not enough for a normal life, women can have an abortion, with the blessing of the law, and he found out. Now comes the question: what if the woman decides not to, and then has a normal baby? And now through this book comes the right answer from the spiritual world, as the zygote already has a soul, complete with the brain as our housing while in our carnal body, and an abortion would just kill us, giving no chance of growing up. I will let your conscience answer this one, but don't let anyone decide but you!

Orson Welles (1938) broadcasted an invasion of Martians near New York City, and everyone ran for cover. Now Hawking has followed in his footsteps with a better job, because his black holes are invading the universe. There is no place to run.

Black holes are so frightening, even the Creator is scared of them. He gave the genius who created it without his permission the option to erase it and to recognize him as the father of all fathers. Souls exist as hope, an eternal medicine, and a reward. He gave him a talking computer, but it could be more, because miracles and phenomena are not only words. I can prove it personally, but behind that must be good behavior.

We never, ever should take fantasy away from children, because they will grow up looking at a colorless world and their adulthood will be hopeless. Hopeless people will be cruel in their stupidities, because man's society has moral law. The spirit world has laws and rules, from which there is no way out. It begins here, just look around, because every pain comes from a cause as it is the effect.

My experiences over eighty years gave me words to fill from hell to eternity, but who is going to read and learn? Picking at each other, pointing fingers, has priority almost at the level of irrationals, where the bigger eats the smaller and so on.

The whole concept of existence is so marvelous; we are surrounded by microscopic matters, all palpitating with life, including our carnal world. Our own body is composed of them. We just manage the mass from our brains; the majority are beneficial, but the harmful ones challenge us, causing sickness. Many times, they win the battle, killing or deforming and the pain gives us doubts of whether there is a Creator, who redeems the well-lit soul.

The microscopic creatures and the microscopic compositions, which science is now going deep into, should be taken as priorities here, because space is a great enterprise. There is more innovation, but first, we must "put the house in order." Not even scientists or "brilliant minds" are crucified without hope, because the dedication here is being put ahead as a priority, claiming they must find the origin of Creation. Meanwhile, death is taking its toll. You are created and not the Creator. The law is yours from soul to spirit, unchangeable for our benefit. As it says, "the Almighty is above us or anything created, in the past, present, and the future." I found out through common sense and not religious books.

I read it in a book presented to me in 1972. It was old and fragmented, and it took me a few days of painful labor to connect its meaning, but it was worth the effort. I woke up in spiritualism after so much negativity in our world. The negative challenges are almost winning the battle, and Brazil is the number-one battlefield for it, as proved by religion and science. The Vatican sent an official representative to the funeral of Francisco Xavier. He died at the age of ninety-two, as a poor, humble man, living behind a legacy of more than 412 books. He was a popular medium in the Brazil spiritualism (Allan Kardec-Espiritismo) movement. They use the process known as a psychograph, as being from one's subconscious source without one's awareness of the content.

Xavier wrote pages and pages with his eyes closed. He worked at the post office in his small hometown, but his legacy gave him the Nobel Prize as a promoter of peace and love, leveling him as a Brazilian Gandhi. Hawking is waiting, passing a half century, for his medal, forgetting the pedestal for it comes from heaven and not black holes.

The human being dedicated to science doesn't know any more what to do to create new names and discoveries to then put them in a shell. Humans begin in the long run on research. Now, they are lost in the medical field. They go to find the composition of the dark matter, not the Universe-Mother. They are trying to take a picture of it, because the pictures of atoms are an atomic blast that will soon take many scientists to a warehouse. The Creator will not see them.

At the end of this book is my e-mail. I answer it, because I am still around. I would like to know about the Higgs boson theory or Higgs particles. It was seriously predicted almost fifty years ago, and everyone now is after it. As two protons collide at the speed of light in the colliding tubes, they could reach God and then what? The money spent on it could have saved their families from the mortal grips of cancer or provided a way to save the planet from the fatal slow but sure destruction. There will be an earth, but without religious leaders and scientists. There will be statues of Hawking saying on their pedestals: "A brilliant mind who spent his lifetime surviving and dedicated his suffering to black holes and promoting a godless universe." Meanwhile, he could have put his effort into finding a solution to end Lou Gehrig's disease. The world would have had made him into another Einstein, or Edison. He could have discovered something like penicillin that would leave us an angel to keep sending microbes running to the black holes of evil—sorry, I mean of Hawking.

Edison did not sleep, averaging less than four hours nightly (after him was my mother and I), and his 1,093 patents are still around for our benefit. No one scared a soul, and he is beyond a "brilliant mind." The Creator keeps those like him on their feet, because illuminated spirits don't have to stay sitting. God never stops, sleeps, or gets tired.

As everything goes cosmological and universal, science goes after the unreachable infinity, while we as humble earthlings, far from the spotlight, feel the ground with our own bodies. Not from far away but between our heads and feet, as our universe on earth, because our life begins right here, when a couple gets together, like the irrationals do. A new world begins, as the universe itself did. It is us being born in a human body, as the perfect DNA multiplies perfectly. The soul, as spirit, has its own eternal fingerprint.

Has anyone looked at it seriously? We are all in our perfect bodies, but of flesh on a frame of bones. It is as weak as it can be, but merciful. We are grateful and holding on to it by our teeth?

Hawking already let everyone know he wants to live, and it is great, but I am ten years older. I am ready to go anytime or be called up ten to twenty years from now, because to die is not to be finished. We are a grandiose spiritual fluid but realistically human, because all the roads point to a Creator. (The Bible says he made us in his image.) It came from few millennia ago. Those were intelligent people. Today, we always have many people around without the showiness of the "brilliant minds." The costumes don't make anyone a saint. The title doesn't prove reality, giving us common sense.

Few express their thoughts like me, because it is like putting your hand inside a viper's nest without protection, but I disclose realities. It hurts those who were unprepared. It educates people for a better future. Hawking openly said God doesn't exist, and no one confronted him. They were afraid of a brilliant mind, but now my brightness comes in. I have polished my soul for eighty years. I am looking at this cosmologist eye to eye, knowing his spark will not obfuscate mine as mine is on a solid rock of a pedestal. He can challenge me anytime on CNN, and as I sit there with my college degrees, I have eighty years of scholarship around 360 degrees, which includes laughs and tears.

I will leave my e-mail but not my brains, because there will be no soul in it. I will leave what is beyond the grave to science, or no science that will find out about the mysterious composition of the dark illuminated matter that fills the Universe-Mother, including the results of the Higgs boson experiment, which will give everyone on earth or beyond now or in the eternal future headaches. This composition has a formula, but it stops with the first one. Between the composition, there are millions of sequences. If you read the book to its end, you will find an explanation that will give you a better understanding.

About the dream ideas of warping time, it is a plastic tubing. The process is going at the speed of light. It beats kindergarten plays, and it must stop there and not announce it as a future realty. Pain and death are real, and the one way into our world is uniting the sperm with the ovum. This should be enough for any mortal on his way to spiritual immortality to stop the BS and go in the direction of saving our planet, not helping give ideas to Hitler as to how to cremate our carnal existence. Atomic bombs did not come from religions but science.

As I mentioned previously, our universe is in an endless aquarium. We are fish floating in it in suspension, as are all the celestial bodies. We are controlled by the touch of a remote control. Without the help of gravity, we would be fragmented into space just like the aimless asteroids.

Time is warped like plastic tubing. We will get to such a day in the file of the past. This was my idea while a genius in kindergarten. My teacher said it was great but not to tell scientists or everyone would label me as a "nut." The future is just a dream because it doesn't exist. The future is today's present time. Now, I realize she didn't go to Oxford or Cambridge Universities. Having become a genius would have given me lots of opportunities to teach realities and not theories in untouchable's dreams.

Later, she went on scholarships to foreign countries for medicine, as she tried to do what we need badly: research how to control diseases, not wasting her short time here. That is what happened to my daughter; she died in her yearly thirties, because of her job in a high-risk contamination laboratory, not bringing us fantasies. She looked for realities to help humanity have a less painful life. My respect was for her memory and the millions who worked hard for a better planet. They are not on a pedestal, qualified as geniuses and "brilliant minds." They worry about fantasies while the salt and fresh water is becoming useless. It is a peril for our existence on earth. The only celestial body we can live on is now moribund.

I have traveled in many third-world countries, not just as a tourist going to fancy places. I went into their small towns and even into the capitals. The stench of the rivers, lakes, or bodies of water has to be experienced to be believed if you are from a first-world country. That includes Brazil and India, where the people throw everything openly into any body of water. The city sewer system goes directly into any water. At Copacabana Beach, or any beach, I can see the floating contamination. When I desire to go to microbe-free sand, I have to fly nine hours to Miami or any clean island of the Caribbean.

My adult grandchildren stopped visiting me in Rio, because they say public toilets are nonexistent, not even in subways, and the millions of coffee houses or small bars are not obliged to have bathrooms for their customers. If a park has one, it is so filthy that people are forced to pee among cars. This is the planet science must be dedicated to helping, because we live in it and not beyond a world not in our reach.

The energy in the Universe-Mother has always existed, but the dark substance could be esoteric atoms, as the base or the skeleton of sustainability. The ironwork in a concrete column and the coverings have innumerable components. Now comes the nightmare; those components are endless, because the smallest divisions come as quarks. They are divided in six parts, but according to "theories or facts," it can be broken down further. This proves space is always to be filled and can be eternally void. Voids tend to be filled, and it doesn't matter how you press it. The atoms will be together, but there is space among them. When the pressing is off, there will be a separation. It's imaginable but accepted because it is eternally perfectly functional before time. We better let it go, because the "nut institutions" are filled to capacity with scientists, religious leaders, and anyone who doesn't take a cruise to steam off.

This is my third cruise this year; maybe I should live on one. Many elderly are doing it, not only for the commodities, but to get away from tensions in a world not logical to many of us. It is a solution as you go on with the intention to write a fast book or just to let it go before it damages your brain and not your mind. It is you as a spirit.

Take a round, steel jar; fill it with small diameter round balls; then with smaller ones and finally with talcum powder sand as fine it can be. Then, put an American aircraft carrier with 550,000 tons, the weightiest floating vessel (more than five Costa cruise ships together), and put it on top of the jar. Look at it with an atomic microscope, and it will be like Swiss cheese. Don't put water on it because it wouldn't help. Please, don't take it too seriously, because you could lose a night's sleep. That is what happened to me.

This is just what human beings will be doing eternally, as souls and spirits, but the Creator doesn't mind. On the contrary, he is flattered to know we are getting smarter. Consequently, spiritual evolution is really going full steam, for a better universe, but morality must be in every sector, as soul or spirit.

As far as we don't deify the Creator with frictions, we will be happy in any circumstance and will receive favors, miracles, and phenomena. That was what always happened to me up until my eighties. I hope it will continue until my last moments, especially now that time is getting closer. The Creator's clock never stops rolling, because he is the energy.

It seems like a Hollywood fantasy movie, but someone was ahead of me and won the Oscar in 1968. He had a vision thirty-three years into the future, especially on technology: *2001: A Space Odyssey.*

It gave the public the fiction as reality, but was it fiction?

I have always said that many of us are angels who have come here to bring technology and also moral ethics. We are disguised as men and women. You don't believe it. I advise you to open your eyes for your benefit. This movie was a success in every country. What caught my attention when I was forty-five years of age was that much of the technology presented in the movies as fiction became reality a few years later. We were prepared to understand it as material evolution, but deep inside is the soul. I was a journalist asking the public for their views and understanding of the story, and the majority did not remark on the final minutes as seniors or the birth of a new life.

The reason is because the astronaut aged in his long space voyage passing into the future and died, but his spirit came back to earth, born again, or reincarnated (back into new flesh as a baby). It is clear in the movies when the astronaut as a young man has the face of a one-hundred-year-old senior. In seconds, he is an embryo in the womb, floating in space. His face kept changing as the seconds passed until it became a beautiful baby's face. He appeared back on earth in a bed, inside an all-white, well-lit bedroom. The baby smiled and gave his first cry ... And the film ends.

Drums played an array with others. The instrument was perfect, as the movies rolled to the end. In the States, Spain, and Brazil, the talk was about this movie, and I know Hawking had nothing to do with the production of it.

Wrong, in my vision, is the reality of the destruction and deaths in wars or bloody revolutions; this misery brings advantages for our planet, because countries do everything around the clock to win the battle, and they put all on the front line, such as research and experimentation on weapons against the "enemy" human beings. Communications and medicine also are continuous in the gruesome, bloody period. It means the vainglory, at the cost of an enemy, as our brothers in relation to the Creation. The suffering is equal on all sides. The question of his presence personally comes, as the choice is ours in free will, to deserve it with our behaviors.

I felt an empty feeling when a teenage girl from a high-class family in a great party at her parents' mansion in Fort Lauderdale, Florida, just asked me right out, just after she had met me, if I could prove there is a God. I am used to this question, and it wasn't a surprise, even from a young person. A high rate of suicide comes from this group, especially from the well-to-do families. I hope Hawking has this statistic on his monitor; if so, I will give him the Nobel Prize he dreams of, but it will include another fifty years in the chair.

I well realize the ones who have it all materially are the ones who give less value to how hard it is to make a living. Because of this, they do not appreciate having the best materially, while others have nothing. This difference is spiritual. The ignorance (ignorance means not knowing about an issue) disregards the beauty and the means of life, as soul or spirit.

The human being who says there is no Creator, because he isn't around in flesh and bones and in a hot sand desert, ignores even looking at a tiny, colorful flower on the dry, scalding sand or a little lizard running for cover under a small rock, at sixty-five degrees in Death Valley. It doesn't need shelter or water, just like the small, delicate bird flying perfectly against a high wind. He gives us all the signs for us to follow, and we are the ones, in our free will, to get lost on our own. We are stubborn. Many of us have so much negativity in our free will that when something is wrong, we just sweep it under the carpet. The question about the Creator is one such issue.

After flying only few hours on my own at the age of sixty-four, I took off in the wee hours on a cold, clear winter night for the first time, defying man's law but not his laws. I pushed the limit from fifteen thousand feet, reaching almost eighteen thousand. Outside, it was about fifty below zero. The air the plane was floating in felt so dense, it was like being in a gel aquarium. I could fly very slowly or even stop. I looked up into the infinite dark matter. I wondered why it wasn't as black as the ocean at night. We can see through it like a gel with a light glow.

The feeling I had as I gazed at it was that I was drifting into it; for a moment, I felt as if I was abandoning earth, and it was marvelous. Suddenly, I got the sensation that I was going to be lost in its immensity, and it gave me the chills.

I observed the warping of earth on the horizon. A tiny line of red surrounded it, and then came the yellow. It began to widen, and the tip of the sun showed up humbly as I stared at it, but in few minutes, he looked at me, saying he was my life. I had to close my eyes, seeking my dark sunglasses, and screamed at the Creator: "God, my Lord! I am here!"

Then I whispered, "Sorry I screamed, but everything is so splendorous. I have the thoughts but not the right words to describe our world. Thank you."

After I descended the one hundred miles to Caldwell at the minimum speed, above millions of souls, where everyone was dreaming and was awakening again to face another day, the sunrise was spreading its rays for our glory, bringing us together, united as one family. We could and can be happier, because in love, there will be no negativity, no wars, and so on.

5 The Universe Has Borders?

As scientist Stephen William Hawking affirms, the universe has frontiers, because there are many universes, as a theory. It is as good as "I believe it could be this or that" or as "speculation." I have felt for years he was making a mockery of people, with his black holes, but he was venerated by a few million admirers as a "brilliant mind." It was mostly the ones at his level with his theories, as he spread them to the four winds. He is not a creator, but after analyzing his abnormal body condition and lifestyle, as atrophied as it could be, his viewpoint on life is a big zero. As everyone agrees, he waits for his final breath, the end of his ordeal in an unfair existence, at least for him. His opinion is God is not ready for him and has kept his body pumping. His only way out is to say: "God, here I am!"

The horizon and free will are available for every rational to use intelligently, even in an odd, unsatisfactory body. The thoughts are free, as Thomas Paine said: "Man can chain another man, but not his soul!"

It is beyond my comprehension how someone whose work is based on theories and assumptions became famous, as almost the only one capable of thinking while the planet right here is fading away in need of solutions from science and not black holes, against which there is no defense.

He affirms he has the time to concentrate on the study of the cosmos, his quantum theories, and so on, because of his sickness, which limits his physical movements. His deepness in science's challenges is his only objective, but he took the path. If he had taken the path of medicine, I bet it would be brain surgery, digging inside the billions of neurons seeking the soul.

This is not a reason to credit him or pity his loveless fantasies, as he is the only one carrying his cross. In his fantasies, he affirms the universe did not have an origin brewing to begin it. He should have to think about our own beginning as a soul in a carnal body, because as human beings, our speck as a sperm meeting an ovum should deserve more attention as to who mastered it. Everything else has crippled his body; there is no effect without a cause, and he more than anyone in science should find that out, because, in his condition, I would be fighting to my last breath to defeat the crippling world of microbes. Not picking wrongly, I would decide to fight to win a miracle and stop those calamities, because nature is too perfect. There would be no reason for us to exist at all.

The more deeply I submerge myself in this subject, the more frustrated I become with those groups. I am even frightened, as we are surrounded by unfeeling brothers and sisters, and it gives us the feeling that evolution or progress could go into reverse, warping time, but it is not possible. That would be unjust.

With those thoughts, it is impossible to appeal to our Creator to help them, because unbelievers don't make miracles. From the emptiness, nothing will come except a grave in which to bury their soulless bodies. The only thing that will come from it will be the flashing fire from the decaying bones.

In his dreams and fantasies, which he has right in front of his monitor, Hawking is credited as a top on cosmologist, quantum theory generator, and the author of the frightening black holes. He has now come up with the affirmation that there is no need for the Universal Creator. His latest book is more senseless, as many of his dreams of materialism never meeting spiritualism were not approved by the notable Einstein.

Did the theory of multiple universes come from a big chicken farm? I am not disrespecting anyone, because as a scientific man, Hawking is the one who came up with the idea we came from an egg. It looks like it is here to stay.

In God's mercy, the logical discovery of a nest as the Universe-Mother has illuminated the dark matter, as the prior eternal existence of everything. It came from the mind of a journalist from a third-world country with endless talents, including piloting as a daredevil seeking God in the clouds and beyond. After more than forty years in the United States, he saw the world, walking, swimming, piloting, enjoying life in first-class and also sitting in the plane's tail. He was learning as he talked to angels and devils. Then, he picked up his *Parker 51* and challenged the godless gang of lost souls in a world of theories that will run facing realities like turkeys at the bang of the first shoot—especially the ones limping on one leg with a broken wing, as Hawking pulls the strings.

The big bang theory is approved by everyone. I asked the public in several countries, and people from all levels of society, from the rich to the unpretentious workers in open fields and assembly-line workers, and they all feel pity, more than sorry, for the one who put black holes into the big bang as a promoter of a godless universe, killing his chance to help medicine and wasting the time he could have spent helping others intellectually. Thousands of disabled people live happily, happy to be created as better times are ahead as spirits. They laugh and enjoy life as it is. (*National Geographic* and Discovery Channel are our best informers.)

Metaphor is the creation of the Vatican (Catholic Church) on the *Humani generis* on which it assumed a neutral position related to evolution. This act takes my thoughts back to history, making me remember *Pontius Pilatus* washing his hands, demonstrating his neutrality related to Jesus's crucifixion. It means if someone tries to stay on safe ground, picking neither side, his staying on creates a big dilemma for the Creator in deciding whether the guilty one is good or bad.

"The Church must survive" gives me the feeling it is the ointment of eternity in any circumstance, and why not? Now, it recognizes Galileo Galilei. After an eternity, it makes no difference, because earth will be continually rotating around the sun with or without the church's inquisitions or blessings, where Satan's match was the "fire of purification." It is lit almost eternally, and people say graciously, as a joke, that it was the charcoal grill of devils.

The Nobel Prize is given in the Pontific Academy of Sciences of the Vatican, which rolls faster than the sanctification to sainthood; their own theories are accepted today, and the competitions are not around, much less the "purification fire to save the soul from hell" or the perpetual reclusions at home. They are more merciful to those like Galileo, because he had wisdom as facts and not "a guessing game called theories."

I am surprised Hawking admires the almost roasted scientist, because his invitation to a meeting in today's Vatican policy could have been for his cremation, because he is not disputing the rotation of the planet but the existence of the Almighty Creator. The temptation is too strong to pass, and it is a good thing he did not stay for the roasting barbecue party.

The science of the Vatican is private, and their opinions, theories, and research are inside the walls. They do invite everyone to come inside and be included. Hawking demonstrated even Godless sinners are welcome, but a guest's views are filled without comments. They are better than "burning wood," because evolution is here to stay.

Now, we know we have many sciences. It is good for the direction of a better planet. They might find solutions as to how third-world countries can keep the water clean, put food in everyone's mouth, pay a fair salary, decrease crime, and—most of all—use the atoms for energy only.

I did copy from a newspaper article the declaration of "Pope Benedict XVI," because the Vatican has a tremendous responsibility on earth today as related to morality. It is seen in an angle based on the love of Christ, a seed they worked so hard on, and now, its fruit belongs to everyone, religious or not. Nature is a book whose history, evolution, writing, and meaning we "read" according to different approaches to the sciences, while all the time proposing the fundamental presence of the author who has wished to reveal himself therein.

The mind is us as an individual. It is untouchable and invisible, because it is our spirit's intelligence. It varies from each one, as no one, because we are not in the same level of intelligence. We now know it. The example is in the family. In eight children, only one was born to get ten in school marks; the others are not even near the five, but they get their graduation, because the school system needs the chairs for the new students. There's nothing wrong with that, because our world is perfect. We need laborers. Who would do the jobs, such as bus driver, chef, housekeeper, assembly-line worker, fruit picker, and the beautiful list goes on and on. The jobs in medicine, engineering, science, and even piloting, and so on are for those classified at eight on a scale of ten. It is always short of them, and they should not brag about it, because it is just a profession. We need the ones with low grades, or there wouldn't be bread on the table.

Well, let's take the special intergalactic ship Star Trek with its speed. We will make our voyage to the limits of the universe for only a few days of pleasure; also, at the borders of another universe or into the Universe-

Mother, where there is no beginning or ending. Hollywood gave us all the necessary instructions on the series Star Trek for TV and movie theaters, guaranteeing it is not only my imagination, because Hawking was the first one to put borders between the universe and the universe's neighbors, as he noticed them in his laptop.

"Mister William, sorry to interrupt your writing, but you are smiling. Yesterday, I looked up amazon.com on the cruise ship computer, and your sister and I ordered two of your books. My English will get better as I read it."

"Please, my lady, can you have coffee or tea with me? Then we can do some talking, while our cruiser is steaming toward the land of the tango, the great Buenos Aires?"

"One thing that has always worried me since my teens is what was before the universe, as we are part of it. Did it begin with a blast from an egg?"

"The only thing I didn't have to worry about because Hawking wasn't born yet. I am eighteen years older than he, and his future black holes and godless existence weren't around to destroy my hopes in my young life as a teenager.

"If we follow Hawking's ideas and theories, the egg exploded without the chicken. If it is true and proved by him, then we are without the number-one fowl, as the most popular meal, which is also cheap, because the discovery of DNA is recent. Please, don't mistake my serious thoughts for tales, because a sense of humor is good to alleviate the burdens of the soul. It keeps you reading the book."

"Senhor William—or do you prefer I call you Bill?—I need your opinion, because I am in doubt about a godless man being famous in science, taking from us the hope of a future life as a spirit, and on top of that, he puts black holes all over the universe, scaring us. Should it be illegal?"

"Sorry, my lady, to be laughing, but it amazes me how the majority of people fall on fictions, because of a lack of education. Don't worry anymore. Anyone with sense, even if illiterate, just needs to look around, stare at the night sky, or better yet, see his child coming up from the uterus and lift the baby up toward the borders beyond earth and shout with respect and joy: 'Thank you, my God, for everything.'"

While tears cascade down his cheeks, the family kneels gratefully. Another miracle has happened from the fountain, as a marvel of Creation.

The dark matter is now a priority in science. It is about time. As a child (around ten, in 1943), I told my father and friends while in Belo Horizonte (MG), Brazil. They had meetings all night on Saturday nights to listen to the news and discuss World War II. I was the only one allowed in.

The BBC out of London was available on shortwave radio, and they heard the bad news about what was happing in Europe, the Middle East, Africa, and Asia that was affecting the whole planet. Brazil, like the United States, finally got involved in it. They considered me a smart young boy.

When my old folks and friends began talking about the universe and science, my eyes used to sparkle. They knew problems were going to come, and they loved it. That time, the black holes were white and invisible. After Hawking was born, they became ripe and changed to black as the telescope advanced in technology.

I told them the dark mass visible at night was more than the whole universe, earth, and all the colossal worlds. They are grains of sand in the infinite ocean of darkness, and everything came and comes from them. Many times, they asked me where was God? My answer was, "He is everywhere to keep his creation tuned."

Food is part of living, and eating gourmet dishes is our privilege, as is everything else the Creator intelligently created for us. That night, aboard the *Grand Holiday*, my choice for an appetizer was duck comfit, lobster tail, and oysters in sauce au gratin; for dessert, I had warm chocolate petite gateau covered with a dense berry sauce and resting on a ball of creamed French vanilla ice cream. My companion selected *ossobuco alla romana* as the main course, but we always shared spoonfuls, tasting the dishes.

Wisely, I put on my list of education, gastronomy. It was a wise choice to be one of my favorite pastimes, as writing, reading, painting, designing, and flying were everything I dreamed of as a hobby, especially when times got tough, requiring serious thoughts and decisions. I ran and still run to the kitchen as my hidden heaven. My grandmother and my mother were gastronomic ladies; they used to invite me to the kitchen, saying I was the only grandson who made complaints about someone's cooking, because I was born an epicurean and the only way was to teach me to be a gourmand. I couldn't have become a genius in cooking if it weren't for them. When they passed away, I was already hooked on the pot and pans, as a family legacy of enjoying one of the great things life offers to us.

In Genesis 1:3, the Bible from the Roman Catholic Church says of the creation of the universe by the Creator: "And let there be light."

"And there was light, because darkness was over the deep."

Common sense, deep in reason through religion, science, prophets, a "brilliant mind," or someone else shouted it on the backseat of a synagogue, and it became a famous sentence as the icon that echoed to the corners of the earth thousands of years ago, affirming there is a Creator. It belongs to every generation eternally, as logical for us to stay happy without seeking further than this, because everything is in it. This exclamation was all he had to say to his sinner's creation, as we build our world in his universe.

As a writer or whatever, without any of today's gadgets, it is just amazing how wonderful our short lives on our great planet are. It is always partly covered by white clouds like cotton balls from heaven. Heaven is beautiful in light blue, showing the first pictures from the moon, as an art of a Supreme Intelligence.

I believe there is an organization so intelligent and powerful it had to have come from someplace in the past, where there was and is a totally irradiated dark matter. What was before it? So times before time, there was nothing on this infinite mass of a so refined composition, which has everything to begin universes. There are no answers now or ever, as soul or spirit. There will be for us eternally a mystery that makes people like Stephen Hawking, millions like him, and me melt our brains just thinking about it, but I have my fan club, the one that believes in the effect of the presentation of the majestic creation. Behind everything in our view and beyond our imagination, he is there and here, as the master of all matter.

I do have a solution. Take a great cruise and enjoy the company of the crowd, packed like sardines in a can. Enjoy conversations with strangers as buddies for a short while, savor the food, and most of all, feel the rocking of the ship and bullshit with everyone while tasting a good, cold drink with some spirits in it like a frozen one. The rest of time watch great shows. It will cool any stress and keep the Creator at ease, because this is the way in and not the way out; otherwise, your life on earth will be a miserable one, as you will be the loser with eyes on the monitor like a harebrain.

But the big bang is without borders, because it keeps expanding where infinity is ready for it. To simplify the notion of it, it is like a pond of water on which we spill ink. This ink keeps expanding, continually pushed by slow currents, coloring the mammoth dark matter mass of infinity with an array of color explosions. It creates new compositions, forming an imaginable world. It is viewed by our eyes through a lens made from melted sand (silica and alkali-like soda).

We can see all the material in our vision and imagination but not where it came from. How about the principal of everything and what was before it? There is no end, but even the most innocent mind has the sensitivity now. Ahead was a steadiness of an eternal pass beyond the extinguishing of a carnal body. That came from nowhere, as the majesty of an endless beginning.

Truly, our only option is to enjoy it, continually, as good people, grateful to be in it as intelligent beings. We see the difference between irrationals and rationals who can take a cruise with all the goods offered and comfortably write a book. You want to know more wisely about what it is all about, if not just to be a happy or a fair person. One day, you will want to belong to those interested to know more. Begin reading and researching, because many books can help us, and culture is the name of the game.

The ones who understand what my writing is about, on material and spiritual comfort, I am grateful to you, because nothing comes easy. Nothing is easy here, but we have hope, because we are not officially citizens of earth. At least that is the automatic process of the death of the material body before we go on to another dimension, which must be better. We cannot die twice, and the spirit is a fluid body. There is no blood to spill or heart to stop, and consequently, there is no doctor's visit.

"Mr. Bill, can I have dinner at your table. I promise not to ask any more questions, but, sir, you are an angel of hope. It's a rare fruit on earth!"

"My lady, tonight, I am not having dinner. Later, I will eat a hot sandwich in the cabin, because I have a lot to read, but early in the morning, I will be at my little table in the buffet area."

To Read Indiscriminately Is to Educate Oneself

If I hadn't supplied my soul and heart with extensive learning from good books, publications, dictionaries, encyclopedias, videos, DVDs, articles, messages, and memoranda, I might be throwing my work into the wind, imagining a lost soul would find it on his way to happiness, understanding there is a Creator. Love will bring us together, as aloneness will be eliminated when love conquers evil. I will rest my head in peace, as I have at least tried to do something useful. My mission is accomplished.

Due to my effort in nearly conquering a place in the sun, I am able to distinguish good from evil and angels from demons, hoping to find glorious axioms from the ones predestined to bring us hopefulness in the face of overpowering negativism from the ones bringing devilish confusion and chaos to people in harsh times, as life is on earth.

I found the history of the one who came here two millennia ago. His life was a short one, as shown in the last three years of his young existence. In his epoch, they had no communications at all beside the sound of drums. Blacks and whites could be slaves; crucifixions and torture were done in public places as a penalty for wrongdoings. The physically strong ones or politically powerful were always the winners. Public entertainment was gladiators or swordfighters slashing each other to death. It was the fascination of society. Women accused of infidelity were stoned to death by men in the square.

He came at the right time, as immorality and anarchy were like the primitive state. Rome and other parts in Asia, the Middle East, Africa, and America were in the uterus, and the power came from the sword. Blood-spilling was glorified.

When he spoke, there was no CNN, radio, or newspapers beyond border lines. He left no written messages and did not even go to grammar school.

His associates were a dozen illiterate fishermen, and no one ever called him a "brilliant mind" or genius, but he knew what was beyond the cosmos without looking at a computer or going to a famous university. He did not receive medals from presidents or scare the public with black holes.

He looked at the infinite illuminated dark matter as the Universe-Mother, lighted with billions of galaxies. He just whispered to be heard: "The house of my father has many mansions."

I interpreted that to mean the mansions are our homes the day we leave our planet as spirits. It makes sense, because it is absolute hope and it is better to die with hope, than looking at the empty, windowless grave where the hopeless incredulous will bury his body and soul eternally.

He said to love each other as ourselves, because of our greediness, and not just our family and few friends. He made it clear that the ones who live by the sword will die by the sword. This is a deep one, because this sword is everything morally. The result will be physical and moral pain, which comes in millions of ways, and it is visible right inside our homes; just look around.

He left us a speech called the Sermon on the Mount, which he spoke to a few hundred, told little stories as parables as easy to remember, and experienced his inglorious death being crucified for rebelling against the high priests, who were without love. The powerful Roman Empire that ruled by the sword backed them up. His followers ran like cowards from the scene. He foresaw all present and future, saying that only with love in any circumstance would we reach heaven.

But he is the only one who had so little, not even a place to rest his head; his possessions were a simple robe, a rope, and sandals. He had wisdom to leave for eternity, proving we are like him, a spirit in a carnal body destined to self-destruct anytime after birth. He stated he wasn't from this world. He meant neither were we, as his brothers and sisters.

He never wrote a sentence or gave theories of the universe or its composition, because he knew from logic that no one on earth can ever reach the stars, much less understand the mystery of creation. We are only able to gaze at it, as far as the telescope will reflect it in our retinas, which receive the images that are at our noses. They are reachable as spirits or in our fantasies, while we wait for our moment to go beyond the death of the body.

His weapon or sword was the right words, the only ones that could reach our consciences, leaving them to everyone, including the ones rebelling against the Creator. His parables gave so much logic as reason for our daily lives that they went beyond religion, borders, and social status. When death is in front of our noses or pain is overwhelming, even in financial loss or at the front line as the bayonet is rushing toward our bodies, the plane is falling wingless, the bombers over our city are beginning unmercifully to drop the destruction of our homes and lives, or at the tragic death of a loved one, as with me and my daughter Carol, when crushed at Times Square—everyone, atheist or religious, will be filling the air with, "Jesus!"

This exclamation is the last resort I hear when a deep problem arrives or when death is imminent. This is true at all levels of society, even while visiting dozens of countries, including in Asia and the Middle East. People of all walks of life, at parties or walking in the streets, hear all the time in conversations, "I am an atheist," or "I never believed in a Creator; that's why I am Hawking's admirer, because this man has enough guts to publically express his feelings." They forget their icon is nailed to a moral cross and there was no way out, except death. He was afraid of it, but he had no choice. Death is real, but eternity is real, because of the continuity of our fluid body. He will know soon, as the aging process terminates any penalty.

"I don't judge so as not to be judged."

"The ones who strike with the sword will die by the sword."

"In giving, I will receive."

"Those without sins should cast the first stone."

Some friends, even the nonreligious ones, as they affirmed, have sometimes looked at a spectacularly clear night encrusted with glistening lights and whispered, "Jesus said the house of his father has many mansions." I just kept quiet knowing millions are just lost souls, puzzled by his existence.

It proves that beyond the galaxies and the eternal celestial explosions are worlds in which we can choose our new address.

He left a legacy but not of having graduated from the best universities; discovering great thoughts to help humanity; being a ruler like a Julius Caesar, a dignitary like the president of the United States, or a pope in the Vatican; or talking on CNN, linked through all TVs and heard by millions. Other religious leaders offer their words for peace or war. It doesn't matter who they were or are; a few months later, their names are gone into the whispers of the wind. Jesus's name is alive in all hearts, because no one ever said one word against him. We write the pronoun with a capital letter for God and never say his name in jokes that don't bring morality to it.

When my daughter Carol died in 1996, in Times Square, killed by a truck, Galil, her husband, a great man from Egypt, a faithful Muslim praising God five times a day for a few minutes, went to the city morgue and looked at what was left of her body. He hit his head on the ground and screamed in agony, "Jesus, help us, because Allah is great!"

Jesus is the only one surpassing time. He became history's number one. He is still alive here in everyone's heart. He did not use materialism to promote love, but spiritualism, because we are spirit and life continues after the death of our material bodies.

Stephen Hawking talks and talks, leading to nowhere, because he uses mathematics, not spiritual common sense. People generally don't care for numbers or calculation but works of consolation.

When love and common sense are above reason, the name will stay here eternally. It does not necessarily mean being a "brilliant mind." Jesus was a simple carpenter in a dusty small town in an occupied little country back in time. He traveled to the cosmos seeking wisdom. He came, looked up, and saw the Creator with the eyes of the spirit. Now, his name is up on a solid rock pedestal. Others are on a sand one, waiting for death, to be washed immediately away by the rolling of the present time into the future.

7 Soon, I Will See You Because God Is Hope

Science is as necessary as religion. It brings us knowledge attained through study and practice and provides an array of definitions. It brings the marvelous world of medicine and technology to us.

The deepness of science doesn't reach the general public, as they are worried about obtaining their daily bread, who's going to win the ball game, and so on. The majority know about the divine egg, which originated the universe with the big bang and scary black holes. Some of them know Hawking as the scientist paralyzed in a wheelchair.

Like in aviation, millions of passengers see the airplanes as a faster means of transport and ignore everything about them, including the origin and the complexity of flying. Prior to taking off, the instructions for emergencies given by the crew are ignored by everyone onboard, and the crew does it in a few seconds, knowing they are not being watched. They do it as a job.

On cruise ships, the regulations are for everyone a half hour before departing to be on the lifeboat deck with their lifesavers, for emergency instructions, and the few who show up go for fun carrying a drink. Religion is like science; people only know the cover of the holy books. Issues of magazines like *Scientific American*, I used to read at the barbershop or dentist; weeks later, they would be put in the garbage because no one would even open them.

In science, the genius in it, as in any field, is the human material progress. Without the evolution of the intelligence of the spirit, there wouldn't be technology or all the marvelousness of the dedication of the few for the pleasure of the general population. It all began when the first members of *Homo gautengensis* (as the latest finding in South Africa) began throwing rocks at each other and eating raw steaks. Evolution took its toll, and time kept rolling. Adam and Eve are becoming part of mythology. The Vatican's scientists are keeping their mouths shut, letting the famous couple rest in peace. It is also included in spiritual evolution. In the last twenty years, I can't recall hearing about the famous pair, even in Brazil. The Vatican says everyone is Catholic, but Mary was promoted as God's mother. It's fine with me, as it takes some pressure off Hawking's godless universe.

Without research, it is impossible to have material progress, which is a combination of spiritual arts. The skeptics have the right to their beliefs or theories but not to interfere with others, throwing their opinions into the four winds and bringing spiritual anarchy, which affects the material world as the effect of a cause, like an inquisition.

My lessons in life were a little of everything and not one altogether. I was eighteen and had good grades in school when my mother asked me which university I was going to. She preferred medicine, because there is always a need for doctors as good professionals. I would be an excellent brain surgeon.

My answer to her was if I began a career in a university or in any extensive business, I would be focusing on just one degree of direction, not the other 359 degrees. I wanted to know a little about everything of life, especially about the question of "why life is like this" and not the monotony of an existence in black and white, without the essence of colors.

I spent a great deal of time in my teens reading and researching everything about life, while the others were playing soccer or sitting in groups talking nonsense. There were countless hours before the sun reached my bedroom window, while I was reading all I could put my hands on, especially about science, astronomy, philosophy, medicine, wars, and music as classical and from the great bands and documentaries. I read about everything from fauna to mechanics in general. Too bad there were no computers yet.

Aviation was and is my passion. I culminated my studies in getting my private pilot's certificate at the age of sixty-four, at Teterboro Airport in New Jersey. I stopped a few years later when I was doing my mostly night landings, seeking my lost soul and staring at the illuminated dark matter, stopping when I did my last and one thousandth landing (1,042 exactly), as I had promised my daughter Carol. It was six months after her tragic death, but I had finished my dangerous dream.

My daily bread always came in generously after serving the call of duty from the Brazilian Army. At the age of nineteen, I began a radio program with jazz and soft music. I conquered my journalist identity and began writing for newspapers and magazines (in Portuguese).

At the age of twenty-one, I decided to improve my English by coming to New York City. It was 1954, and I met a young lady who stole my heart. I was staying and decided to open a Brazilian restaurant near Rockefeller Center. It was an absolute success. I had been born with a pen in my right hand and a golden spoon in my left one. A few months later, the Brazilian Club Restaurant at 150 West Forty-Ninth Street was appraised by the *New York Times* on a Friday with three stars, with a comment there should be more places like mine. It opened my doors to New York society, and having consulates—such as the Brazilian, Portuguese, French, Spanish, Australian, Irish, British, and so on—within reach of my charismatic golden spoon, I and my spouse never missed parties or wine tasting.

Nothing in this world comes free, but with each of our efforts, as the saying goes, "What we sow, we must harvest." To stare eternally at the colorful, infinite lights in ecstasy, dreaming a shooting golden star will fall into your backyard is like buying a lottery ticket and planning what you will do with the millions you are going to get as the winner. Staring into the cosmos or at a saint's image will not transform us into "brilliant minds," and dreaming is not enough if we do not pull up our sleeves to make it a reality. We must elevate our spirits to the level of the question of "why life is like this." We must discover that in the aquarium of the Universe-Mother exists all life from the smallest fish to the blue whale.

When we research individually and in general all the facts and things, to gain knowledge of life from A to Z, it will bring us moral, material, and spiritual comfort, but we distance our souls from the general public, waiting for them to educate themselves and to use our free will. While time rolls on, the present slowly moves toward tomorrow. An example is the big bang explosion. The entire planet's existence is due to the egg, which is in all the tables, but without the chicken, some lost souls are orphan eggs.

Saying one does not believe in a spirit or the Creator is illogical to me.

If a person doesn't educate him- or herself in the 360 degrees of knowledge or at least a few more than one, his or her mind will be begging to open the windows for more vision, to upgrade the IQ or to become more capable of generating ideas.

It is just beyond my thoughts about the ones going full blast and rebutting the Supreme Intelligence, as being completely obfuscated by the beauty of our universe, while breathing and feeling the magnificence of our existence, which begins with the creation of our carnal body, as a sequence from two elements: a man and a woman. It should be more than enough to imagine the magnitude of a being, a master of peace, whom we call God.

Sincerely, when someone says that everything comes from a void and death is the extermination of life, now I just walk away and keep my distance, because they only see the world like chickens staring at the gates of the barn and telling the chicks it's the end of the world. The view doesn't pass the top of the mountain of the carnal world, but the mind is continually growing beyond the levels of the grave, because those are the rules of evolution.

"Why should society call those people "brilliant minds"?

Not everyone is able to look and feel the deepness of others' minds, and the average person puts pity first and uses blind reason, because blind faith is our worst enemy, as stated by Thomas Paine.

Yes, it just now came to my mind that Hawking became a "brilliant mind" because he rebutted the Creator. It is for that reason I am writing this book, which could even give him more glory, as everybody falls into his trap, knowing that a terrible-looking man, totally paralyzed, has more intelligence than all of us put together. It is his right to say, but to most, the consolation for a suffering humanity is the hope of an eternal life, even a paradise. It goes as an afterlife. Is it bullshit?

It is also my or your right to counterattack his malicious thoughts as a revenge for all of us. He can't get up and walk or have a normal life, while he's bitterly mummified, almost an eyesore, as a human, with or without a brilliant mind. As for me, he's not, because his world is counting black holes, dreaming of quantum theory. It doesn't matter how many honors or medals he gets; he can't even hang them on his neck, embrace a woman though he says he loves them, or go to the bathroom as a private act. He keeps writing and saying, inclusively, that his condition is not a curse from God, but just nature's law.

The reason for worrying about his condition in public is as he aged, his looks became so irregular he can feel the misfortune of a life of negativism. No one would be proud to have him at the dinner table, but his conscience is not bothering him related to the Creator, because of a God who gave him fame and by the same token paralyzed his ego, but not his tongue. He became the Creator's hobo.

It happens to those who go around proclaiming to all the winds, the trade winds—the media—arrogantly, like Satan as Hitler was called, or an anti-Christ, and they reincarnate on earth disguised as high religious leaders. Their actions and tongues are the revelations of the soul.

It reminds me of when I was visiting my brother in San Diego, California. I rented a Cessna Cardinal RG and decided we should go to Las Vegas. My brother, his older friend, and I crossed a cut in the Rockies, and in minutes, I saw the endless desert. In the distance, it looked as if it was only a flat, sandy world like the moon, where I could trespass its horizon and be lost in space. It was scary. This trip made me realize we do things without realizing the consequences. As we were crossing the Rockies in a small aircraft, the wind was blowing us a few feet closer to the canyon walls, while my passengers looked on with happiness at the amazing scenery. I came back two days later during the wee hours for a calm crossing to find out we did not become another small-plane statistic, because I had to write some books.

I signed, bought, and picked up at the library books of all kinds, but in the line to educate me, not others, because to know is as personal as our own soul. The past is part of the present and means more than the ones who say yesterday is old news and we should not bother about it now.

Nazism, representing Satan disguised as Hitler and his associates, almost destroyed the planet, killing at least 240 million and injuring billions (estimate) souls and rendering others homeless. It took years to rebuild their homes. Tokyo became their evil partner. When it was over, I was twelve years old and knew nothing at all about the conflict, but twenty years later, I knew all the details about all the wars, including those of Korea and Vietnam.

Many relatives and friends used to say (they are all now in heaven) that I was obsessed with knowing things that would give me values, but they had all had a mediocre life, not enjoying the beauty of it, happy to eat rice and beans every day and canoe on the small city lake, like taking a cruise.

In the big bang explosion, there came the little chick, which we call the universe. It is a name we created, just like God, atoms, Joe, and Mary. Now, the big nest will make the little chick very happy, knowing it has an origin, the Universe-Mother.

The not-forgotten past is printed and recorded in the library of time, providing lessons so we do not repeat errors, so we learn from it, to make the present and the future better for all.

Life is not for sitting around talking and not getting involved in the problems, but not doing it with a remote control, staring at a monitor in a room, like many politicians, dictators, kings, or scientists, not feeling real life. This is not doing a good job, but anarchy, like Hawking, in his chair doing three miles per hour at altitude zero, is unable to see the world, traveling and talking to souls, feeling their problems, eating exotic food, facing nature's beauty. I have suffered moral and physical pain, almost hopeless in situations, seeing death face-to-face and worrying about my daily bread. I have seen hundreds of family members, friends, and strangers die, and so on, on and on. He can then open his ungrateful mouth, as even he must be fed and be taken care of at the cost of hardworking taxpayers, as he takes people's hope. He feels life is hopeless, but what he did to deserve being chained to a chair is right in his conscience. If not, he is just a cheap blasphemer.

Over the ten days cruising last December, I wrote this book. I had to control the writing, because the majority of readers take too long to read and even get bored, even when they are interested in the subject. I was among more than 2,600 souls. We came from all walks of life. I was researching in the spiritual field. I can't be away from people when I am doing my writing, because there wouldn't be any inspiration. I would have gone to the pits, because everything I do is related to us, and the stimulus comes from facing the realities. My subjects are part of it, here or in the galaxy. With those motives, my pen goes nonstop, giving me more desire to do my job, not as a novel or an equation calculator.

Sometimes, like everyone else, thoughts come to my mind. They are not dreams or fantasies but based in events, facts, and realities, as if I could have had the opportunity to know something big enough to destroy our life in the universe. It was not a way out. I never, never, but never would spread the bad news, because it would be a Satanic act.

When Hawking said he admired Einstein, I wished he had followed his spiritual feelings, especially to the real "designer and builder of the universe," as all the material for the universe came from the Universe-Mother, the dark mass, or even the big nest. The expression "big bang" was digested by the public.

His pride is in his theories (fairy tales). I researched for years. The general public that knows about it doesn't even comment because of his deformity, as if in holy respect. The negativity of it is untrue; it gives no hope to anyone. He was not searching spiritually for a God, who goes around saying, "One, two, and three," and points to him to have a terrible body in life, but if Hawking had read spirituality to find out we do not belong to the earth life eternally, a light would have reached him as a "brilliant spirit" and confirmation would have arrived that his illness had a purpose too deep in the medical field and not in cosmology, saving him the "big nail."

Cosmology and quantum theory gave him medals but not a Nobel Prize, as nothing from him can save the planet or improve medicine, industry, and so on. There is only the despair of a world without hope. If he had, for curiosity, read the Catholic Bible, as I did in a few nights, it would have stopped his big mouth. I am thankful for the religion. They put in their bodies and souls for it to be the number-one book ever sold, as in the words of Christ, "His blindness was not his or the parents' fault, but the glory of God, and because they did not blaspheme the Creator, I healed him."

If this genius would take one hour a day to research people and problems, his way of looking at life would change. The number of teen suicides is just growing. Alcoholism and drug addiction are now uncontrollable, and divorce is passing the red mark, leaving millions of women alone with the children. The Vatican now worries about the future of the family, as the "brilliant mind" just keeps his retina on the monitor, while bringing negativism that keeps him in a harsh life. It is only his fault, acting in his free will.

As a reward to us from the Almighty Creator, dreams let the soul leave the body and travel freely to anywhere in fantasy, helping mostly those in prison, away in remote places, confined to wheelchairs or beds to reach the deepness of the cosmos.

The ones in normal lives even daydream while walking or with closed eyes on public transportation, or anywhere. Even flying, I had met angels. I am not sure if it was or was not real, but it looked as real as it could be.

The fantasy Hitler had for a perfect human race and to dominate the planet, as he said was his dream, was destructive, and the results were fatal to him and the others involved and a disgrace for the world.

But dreaming beautiful dreams and fantasies keeps us young and happy as we do good for others. I believe I am an example of it. I am not a senior now, but as my grandchildren say, "Grandpa became a child, as he tries to change the planet for the better, but his time is running out."

8 Without Hope, Life Isn't Worth Living

To be a child is to live in a fantasy world of cheerfulness, with parents, relatives, and friends, while waiting for Christmas and for the great surprise gifts from Santa Claus or Papa Noel.

Suddenly, this dream fantasy vanishes in confronting the death of someone dear, especially one's mother or father. How can I write so deeply about those feelings? Is it because I went through it when my daughter, Carol, died tragically and left her son, Christopher, an eight-year-old boy, and we raised him? (From this tragedy, I wrote *Life Is Beautiful Doesn't Matter What, because We Are God's Children*, 2004).

Parents, friends, and even the school tried to cover the tragedy until the funeral a week later, because they affirmed bad news can wait. It seemed like a merciful act, at least holding it back for a short while. When we take a dream from a child, his world fragments and becomes colorless. His hope of having a normal life is usually gone, affecting him to his last day.

The same happens when a leader of multitudes—a president, a sports hero, a scientist, an actor, a singer, or whatever—uses his celebrity status to do evil; he takes dreams and illusions as icons, damaging minds in every way, giving negatives in their behavior and theories, destroying or damaging hopes as the sentiments on earth. Souls are holding on, because we are always children.

I have the right to see someone's errors, not in the name of the Creator but of an existence where there is no morality, because without it, there would be a void of a Supreme Intelligence and no one to motorize his creation. Anarchy would rein, and there would be no functioning existence without an absolute precision or the sentiment of love.

The genius admitted that the egg shattered in a diabolic blast; as a theory or affirmation by atheists, any creation is possible without a creator, including the diabolical in the form of the free will, where they doubt an existence in the dark composition, known by science as dark matter, better explained now as the Universe-Mother. They believe it is inexistent; it is seen as diabolical by them.

This black matter is more defined as a substance beyond any idea, even as a composition of atoms or any related material. It gives sense to a science lost in its complexities and the marvelousness of an imaginable ingredient where there is no beginning or ending. There is a perfect neutrality, not interfering with any matter composition, because it is where all existence is held in its bounty.

It can be felt as deepness and absence of light, but it is clear enough to be felt as enlightening by its own radiance. As I said before, the words we have for communications are now great, but not great enough to be a way of showing the explanation as a real picture we are looking at.

That's why those unbelievers should leave the monitors and the pictures of heaven and go up to the Himalayas or cross the Rocky Mountains or the Cordilleras on a winter night in a small Cessna, flying slowly and staring up to infinity. They will have contact with their minds and spirits, landing to then stop spreading ignorance and be cheerful to be part of it.

Illogically, wrongly, many scientists believe the dark matter was created at the moment of the big bang, associating with Hawking and indirectly admitting the origin from an emptiness where no matter existed before. This is impossible with all the marvelousness of this instant creation and also leaves no room for a Creator. What are scientists anyhow? If you look at it this way, I am also a scientist, so are philosophers, cosmologists, and so on, but a good local medical doctor has more value to the general public than all the theories and philosophies together. The title *scientist* feels good, as a synonym of grandeur. In third-world countries, the poor people will bow their heads and call you "doctor" because of their illiteracy.

Without the existence of the Universe-Mother and the radiant dark matter, the universe we are in could even be expanding, because its fragments originated from the big bang. It could not be going to a nonexistent space, as the fish falling from the aquarium will land on his tail.

It is confirmed as reality, because the average human being has common sense, and from this concept and analysis, great discoveries come from minds ahead in their logic, and they all like Einstein. He let everyone know that without a spiritual Supreme Intelligence, nothing could exist.

They study the minor details of everything, from the colossal floating celestial bodies to microscopic organisms and the movement of matter as a rock formation. It was discovered a few years ago by an Australian rock analyst. Surprisingly to all, rocks also move and grow or expand. She left some in a dish under a microscope for weeks while she went on vacation, a fragment to be examined in her job as an analyst for a mine. She took a picture of it before and after, and when comparing the photos, she noticed the differences.

In the beginning, in the scientific associations, doubts reigned. It was considered so illogical that no one even tried to research it, but as the founder of this phenomenon, she persisted. Years later, it was then confirmed; everything existing is pulsing continually.

It was confirmed by human intelligence, as the majority of us agree unconditionally, the whole concept of our existence, the precision of it all, has a designer calculating, and forming everything. Where it came from is an eternal mystery. The omnipotent spirit is fluid. We are souls in our carnal bodies, as previously in the beginning of our human bodies as zygotes. Through research done in gametes, we see the DNA, but not the energy of our souls. It moves it. It is what we are and is of our body as men or women. The fusion of sperm and ovum is the same principle as the origin of the big bang egg: we came from our sperm and ovum suppliers; the big bang came from the Universe-Mother. I agree with Hawking that there could be more than one universe, even an endless number of universes, as we have galaxies. So what? Science must use all the energy to stop warmongers, for a peaceful planet, as a priority. They must concentrate on the issues that are pushing the planet to an end, such as natural disasters, that are at our doorstep.

I believe now, in our present time, the big bang and the black holes are cannibals eating galaxies. They are now competing with Christ's parables, the Ten Commandments, Buddha's sayings, or the beautiful exclamations of the Koran praising God. The only serious negativity is Hawking, who continues blasting about an absent Creator, but his physical condition is understandable, as he can't enjoy anything besides being grateful for being alive. Even in his body mummification, he does a lot with his theories, dreams, and fantasy, as experienced in a material world, like everyone, though, his deeds are not forever.

At least when someone has a stark life because of a debilitated body, the aggression of unconformity against the Creator is normal, and one weapon is to defame his existence, as everyone knows of his absence, physically and vocally. But the ones who accept it have a happier life as a reward, as we see on our daily life or in the great programs like on Discovery and National Geographic channels and many others.

Scientists confirm there is no effect without cause. Those in nonconformity, who feel discriminated against, should try to find out, not directly in contact with the Creator, but through other approaches, how to change their way of looking at the ones living with incredible deficiencies, having a normal life.

Words can be deeds, because they can help or destroy, bringing happiness or anarchy. Big leaders and heads of countries lately have been expressing the negative words of those on the wrong track, such as Hitler, Stalin, Mussolini, Gaddafi, Hussein, and the list from the present and back into history. They demonstrate how words bring war and peace and can make society a heaven or hell.

Orson Welles (1915–1985) was a radioman and actor. He transmitted *The War of the Worlds* in 1938. The news of an invasion of Martians in a small town near New York City (I was five years old) was just fiction, but was taken by the population as serious. The army even responded to save the citizens, as they panicked.

Now, I will state that if Stephen Hawking, with all his followers, says the black holes are getting suspiciously closer to our Milky Way, I can assure *The War of the Worlds* will be just the appetizer of the pandemonium, because it will be beyond control. Everyone will run to churches or any religious institutions and will include all the Almighty's armed forces. When the joke is then explained, it will be too late. It will include the funny godless souls. They play with the public negativity, and the result can be disastrous, because panic is almost impossible to control.

The most interesting marvel is that *we are going to harvest what we sow* because of our free will. Our free will gives us the right even to doubt the Creator but also gives us the right to defend him. I am not dead yet. I am in my eighties, but I have the capacity to get up at 6:00 a.m. (after having gone to bed sometimes at 2:00 or 3:00 a.m.), enjoy two or three eggs cooked in many ways as I prepare my gourmet breakfast, and then as always, I have my own prayer of thankfulness. I do things the young ones envy, from physical to intellectual feats. If I sit, it is because I am enjoying eating, writing, reading, or working on the computer, as I have been now for days, reviewing this book. I continue even with puffed-up feet and hands from the mouse, but I never give up, because that would be a defeat. When we soften, nothing will be accomplished.

My labor is an expression of my thankfulness to the Creator. I am, was, and will be forever, in the present as in the past, as time keeps rolling into the future, his number-one admirer, not particularly through my deeds, as considered by friends. This is based on the reality of how active I am, getting my private pilot certificate and taking them on the planes over congested areas, such as New York City, Philadelphia, California, and Miami, not as to defy death or the system, but as a great senior pilot in age and experience coming for common sense and reason, as stated by Thomas Paine, one of the greatest "brilliant minds." His name is here to stay, as he helped and not scared society.

I always hang onto hope in any adversity. When I faced the death of my daughter in 1996, I was asked by many how I looked at God who had given us an angel and then shredded her body.

My response was when I confronted the horror of my life, I fell on my knees and said, "Oh, God, you are almighty and the creator; you gave and you took, not only from me now, but all of us, in the past and present. You gave Carol to us to raise with love, and she became a lovely young and beautiful lady. Suddenly, you yanked her from our arms, as you do to one child's parents and so on. Oh, Lord, when my time comes to go from soul to spirit, I will then understand wholly the reasons, because everything is so extraordinary and does not belong to us materially and then comes hope, giving to our souls the balm that we will all meet again in the future; that is tomorrow. I love you, my Lord. Help me and my family and friends over the next few days. Amen."

Amazingly, even the ones who proudly claim to be atheists embraced me, whispering into my ear that God knew what he was doing, as we have no choice.

Into my years, as it rolled and kept rolling, at the writing of this book, I am great and thankful not only to the Creator but also to the ones who, in the daily moments of life, give us the royal treatment as love. I noticed the humble people from the hills and farms, laboring for us to have food on our table and the famous magnates conducting great business, moving things and creating many jobs, each one doing what needs to be done in our society for everyone's benefit. Each one of us does what is necessary, as in the beehive. We all are necessary to keep the wheel rolling, as in the cruise ship, every crew member must be accounted for or it could not sail.

If the big bang egg got hotter and hotter as did the sun in the billions of centigrade and in a trillionth of a second, it blew up into an infinite number of pieces according to a man, creating the dark matter from the void, it would be the absence of anything—no matter, no atoms—because there was nothing in existence.

Even illiterates in third-world countries or beggars born on the sidewalk in India, Brazil, or other such places offer a smile when confronted with this question, saying that from nothing only nothing comes. A dry faucet connected to a dry tube will never flow with water. The illustrated will look at you and just walk away, thinking this is another unintelligent human being, a reasonless person.

It is like saying the chick doesn't have a mother-chicken; it was his origin with or without feathers, as I was saying, as an orphan, but all the orphans have a mother, as the big bang originated from the Universe-Mother. Stephen Hawking and his black holes originated from the United Kingdom's universities as a godless robot, because his metallic voice is from an unsent mentalist human hobo, as he acts like one, not upsetting any feelings of his fellow earthlings. Hope is not computerized in his system, because there is no sentiment felt or he wouldn't be openly stating there is no need at all of a spiritual being to create anything, giving the feeling he came from a computer assembly line.

We humans, like irrational animals, have a mother. We come from an ovum, making it impossible to be motherless, but only orphans. The same is our total creation from the dark matter as our mother. I believed

I am pounding it in, but it is necessary for everyone's benefit, because I don't have many years ahead, as a teacher.

Believe it or not, the big intelligence and bright mind is warning us about the destruction of earth by nuclear war or an asteroid hitting us, and if this doesn't happen, then there will be an attack by *aliens*. We should stay away from them, because they are so advanced technologically that there wouldn't be a counterpart to the power of those advanced civilizations, like the Indians when Columbus came to America or the Portuguese to Brazil.

Everything points to him being an alien. In the movie *War of the Worlds* (1951), earth was invaded by flying saucers with long, snakelike necks blasting everything, and there was no protection from our weapons. When we thought earth had been destroyed, they began dying from our diseases, as earth's microbes invaded their unprotected skins from our contaminated environment. Now, maybe I am pushing the envelope, as they say in aviation. We are risking our lives by flying dangerously, but like the aliens in the *War of the Worlds*, Hawking is also being curbed by a microbial power restricting him from doing more damage to our hope.

Now, our "brilliant mind" gets contaminated to a crippling stage but has held on for a half century, which is considered impossible for a human being. He doesn't care about our feelings, blasting our hope for an afterlife. He doesn't bother at all about our medical field; he just stares into infinity, waiting for the squadron of his alien ships to reach us. He is preparing us to surrender without a shot, and his purpose in being so senseless about our sentiments as human beings is now *in the open in my fifth year of following his path.*

He talks about the cosmos, as the only authority on the Milky Way. His voice betrays him to the ones now ahead of his charges. He considers his mind above others, even being restricted by our microscopic defenders. He is trying desperately to find them in his mathematical research into quantum theories. He is on his way to discovering, going down to the invisible world of germs, against which he has no defense. Practically, none of us do, but his vulnerability helped us; otherwise, he could have overpowered our system.

He demonstrates very well that he is an expert about aliens. It is one way to have us scared, as there is no defense against his black holes. He is so intensely an expert on black holes, as his lost world. He is seeking celestial bodies on which to begin new life. Others sought earth, thinking it could be it, but they didn't know about our biological negativity in the form of deadly microbes. They could be responsible for our own destruction. Plagues and outbreaks make no distinctions, because blood is blood.

A godless protection of our conscience would make things easier for aliens, and that is the aim of Hawking, as he authoritatively pushes us toward disconsolation, as there is no hope of a Creator. He is brainwashing us into an easy surrender, as he now affirms that despite his physical condition he wants to continue living until help arrives. Now, we all will keep one eye on him and the other on the sky, while having the UN Armed Forces on alert, ready with our navy's Gatling guns.

This book is my promise, overdue by more than fifty years. I was thirty-one and Hawking twenty-one. I was getting more data and waiting for the right time, as it keeps rolling continually in the present and into the future. I was living history behind open doors, penetrating the future, and the years passed. I never lost hope, because hope cannot be lost, and suddenly, the lightning strikes twice in the same place. I read *The Grand Designer* right after it rolled off the presses. I was visiting Pennsylvania from Brazil to attend my sister Helena's funeral; she was two years older than I.

I bought the book at Barnes and Noble, paying over thirty-five dollars for the thin book. It was as hot as the burning sun; people lined up to get it, but I did get one. I hoped to see if Hawking was out of his inferno of life and had changed to a Creator's universe.

I asked the dozens in line the same questions, and the God ones expected the author had learned his lesson, while he expended a lifetime wasted as a mummy. I walked to a café, and after a few mugs and a great sandwich, I finished it. I got up with a swollen butt, went straight to the garbage bin, and trashed my first book in almost eighty years on our planet. Hawking keeps himself an orphan, but he also has nowhere to run if he could, because he also killed philosophy.

To sit a half century in a chair without movement and voice, I would have had to receive from the Creator a merciful death, which would take my soul then as a spirit back to where we came from. We would be off the chair or in it but with movement, enjoying life like millions of others in wheelchairs. They are even competing in games; they are happily married and have a normal life. That is all we ask our Creator.

But it is like being in prison for a reason, and when inside, the prisoner rebels against the system. His punishment increases. The effect is clear when we go against the tide, something illogical. Even if our feelings are negative, holding our tongues will bring more good than bad when the terrain we are walking is dark and full of black holes, because for every friend we make, we create a thousand enemies. Negative thoughts also frighten or even kill, just like prayer will lead to somewhere.

Consult (amazon.com or any site) for my book about miracles and phenomena. It is very realistic with photos of the most incredible facts ever recorded but forced and ignored. The average human being is still too far away to notice the spiritual world by which we are surrounded. We will be in it anytime at the death of the material body. See *Dr. Fritz: The Phenomenon of the Millennium* (440 pages, 160 photos, 2002).

I have so far received almost a dozen miracles or phenomena since I was a child. Most recently, cataracts were discovered over five years ago, and I was told to have immediate surgery, but at the moment, my vision is normal. The cataracts are even turning blue. My eyes are light brown, and the new eyeglasses are just for reading. I am very thankful. One day, everything will break down, because we all must return to heaven. That sounds good.

There are sicknesses without cures, like many forms of cancer, and millions of deaths are caused by microbes not reached by penicillin or any other medicine available; meanwhile, we see family and friends die defenseless, and motor neuron disease attacks the spinal column. It is very rare but immobilized the genius of cosmology, surrounded by all the best medical care available in a first-world country. It is an example to us that when there is a cure, there is more involved than materialism. Hawking should get his Nobel Prize for surviving a deadly disease for a half century. As someone said, it is because God is keeping him alive as an example to all mankind of someone he is not willing to call back. It is a lesson to Hawking and those on his level to behave.

Almost all theories are scapegoats for mysteries, or better yet, there are no solutions for the impossible questions in any field. One thing I have noticed for years is that as we age, some minds become sharper or the IQ gets better. I am one of this species. And why not? Everything has merit.

That's why our gadgets or electronics from science are finding marvels—computers, lasers, cell phones, and so on, and it will get better to a point. Our bodies, while we are in them naturally, will be transported to anywhere materializing in seconds, as our transportation, but it will be ready thousands of years in the future, and having nothing to do directly with the spiritual dimensions, it is just material evolution as natural.

I know Hollywood uses it in movies, but it will be true. One thing is for sure, men (and today women are also part of it) will never conquer death and the secret of birth from the sperm to the ovum. How everything began will be an eternal question; the infinite future is time, because we have the material and the intelligence to build it, but we are only the labor and not the supplier.

There will never be a perfect society on earth, because we are souls, and our spirit will not have the influence of the planet as a school, and our free will is subject to the pressure of it. The separation of the haves and have-nots and the indifference between individuals with prestige and their families will always separate them from others. It appears to be discrimination but is not, because social status is part of human life on earth. Having good intentions is great, but I feel the difference because I am between both levels and notice that in Brazil, the States, or other countries, as I navigate among both sides, that money is an issue, but money without culture in all levels is not enough. Education is necessary, but it doesn't open Satan's doors on earth.

That's why flying an amphibious plane I could land on any surface. Better yet is the helicopter, because you could come from the roof, avoiding the doorman and the butler.

No one in his right mind wants to die, even when aging comes calling. (Naturally, suicides are not included; that's another issue.) Down deep, no one wants to come back. That is why the idea of reincarnation is not included in religions, as before, except the *Spiritism Doctrine* (Allan Kardec 1869, France), and the adepts are the minority of all religions or cults, not even making a dent on the surface.

I created my opinion, because I didn't see or hear the expression "why is life like this?" and when I mention it, eyebrows lift up, because this question is in almost everyone's heart, as life should be their way. It means there is no pain, no death, no fights, no stealing, and everyone has a good salary; there is no unemployment, and people work twenty hours weekly. You have good kids, and there is no crime, no infidelity, and no planes falling from the sky. Ships wouldn't sink, and no asteroids would hit earth. I am stopping here; otherwise, I would need a hundred more pages for the list.

It came to my mind that I had always thought, since my teenage years, that we can be one of the millions or infinite number of universes as theorized in Hawking's work. There are great numbers of galaxies with suns and their planets, and it is infinite, the dark matter, which supports any volume created, as if we throw a grain of sand in the ocean.

We were enclosed as souls in the solar system and surrounded by galaxies and innumerous nebulae in formations. This is our notion of the frontier, where the limits are. It is impossible to distinguish any borders, if there are any. I watched the fireworks on Copacabana Beach at midnight on New Year's.

Dark matter is the absence of a density of light and is not black because there is a faint dimness that set it alight, a luminosity of deep. We do not see it as a total darkness. I have been in some caves, and when the lights were turned off for a minute, I felt I was not in a 100 percent darkness, though there was a sensation of not being in light, giving the comfort of hope. It is like a child or even an adult putting nightlights in the house that give a glow of the comfort of not being alone in the universe, and it is the mastering of a super power.

The dark mass is the existence, and it's difficult or impossible to explain, but it is easier to comprehend mentally, as words can go only so far.

You can put words onto paper, trying to describe something. The ones who do not see it will have some idea about it, though not the real meaning. Here again are the sayings, "You will feel it when you see it," or "You must see the Grand Canyon to feel its grandeur," and now I add, "At night, staring at the infinite darkness, we can feel its deepness."

In the dark matter, which has some light, it twinkles. When God created the bright light in front of our retina, he just condensed it from the dark matter, because it is part of it and we can detect it in a desert night away from city lights or in a Cessna at fifteen thousand feet up. It includes earth's curvature.

In the dark matter is the storeroom of the Creator, he compresses the material into an ostrich egg but was erroneously mistaken as a chicken, because it equals twenty-five regular eggs. I have been asked by the ones not interested in science how scientists will reach the borders of the universe, and my answer is as simple as two plus two equals four. It is the day they die, and as a spirit, they can transport themselves everywhere in the universe and do so instantaneously at the speed of imagination. As you think of the place, close your eyes (the spirit ones), open them, and bingo! You are there, and this phenomenon is called "transference or transport." The movies and TV series like Star Trek explain it in colors.

When something is inexplicable in its grandeur, common sense and reason admit there is a Supreme Intelligence prior to those phenomena, as the dark matter and the universe or universes are the effect of the cause.

The ones who feel vain unwisely come up with that nonsense; that, to me, is being used to show off above others in their inferiority complex.

Those who are in doubt even in confronting such grandeur must be in a desperate position with their mental problems, such as paranoia, depression, and so on, motivated perhaps by moral difficulties, physical deficiency, or even financial problems. Their scapegoat is totally against God, stating his inexistence because no one saw or heard him, but they mistake this for God. He is so powerful that he doesn't have to appear, because any existence must come from him, on his terms.

I have met people from "high levels" of society, people who are financially successful. They tell me how difficult even going to exclusive restaurants or clubs is. They must keep their distance from people, because they come to them like ants on sugar, trying to get autographs, kissing their hands, asking for favors, and just staring or smiling, making them sorry for their success. They take photographs, keep popping flashes in their faces, so they can't have normal lives like other human beings.

In my successful restaurant, security for the "rich and famous" would come in for reservations, saying they wanted a corner table, but if someone from the staff or a customer would approach their table, they would just get up and leave the place without paying the bill, never coming back.

I would tell my personnel if anyone ever bothered the guest, I would personally throw them out for good. I used to tell the same to my regular customers.

Lady Diana died in a car crash while speeding off from hundreds of photographers, who were chasing her in the wee hours of a foggy morning in Paris. Imagine God showing to his mortal creation that he would then be able to take care of his infinite creation.

On the fourth night aboard, at my little table, in the wee hours, I was looking at the moon throwing its reflections across the ocean. It seemed to be dancing when embraced by the small waves. Then, a very old lady with difficulties pushed a chair close to me. She said had been a widow for the last eighteen years and they had always taken cruises, because they loved it. Now, she's ninety-one, and as long as she can move, she will be cruising. She wanted to say a few words. I told her she had a half hour to do so, while I offered her a cup of hot tea.

"Mr. Bill, the dark matter and the universe or universes make no difference. Everything is God's creation, and that's final. Some do not see the magnitude of creation because they are blind in their ignorance, ignoring beauty and greatness. I must remember 'one swallow does not make a spring,' but it can disturb the migratory system. I should not allow it to disturb the way I am."

While she was talking, I was writing, and then I noticed she had left as quietly as she had arrived. At coffee time, I asked the crew about Ana Paula, the lovely older senior, and none could account her for. Later, at the information desk, I discovered there wasn't any such name on the list either. That night, staring at the moon, I whispered a sentence to Ana Paula, thanking her spirit for having noticed me and my writing. I would never forget her visit, bringing hope of a life after death and giving support for me being what I am.

All the time while writing, I had been worrying about the critics, those who had been hurt morally by my realities. I did not intend to offend but to alert, and it hurts only the sensitive ones. Their consciences bother them when the hat fits the head.

Aristotle is respected as the father of science and philosophy. Millennia are rolling by, as time never stops. It is still alive in our minds, and when I saw what Stephen Hawking stated, negatively, about no longer needing philosophy, I noticed his reputation was going into the quicksand, not only with the general public but also his followers, as I spoke to some passengers and looked at the comments about his last book on the Internet. It means, when the fire is still glowing, we must keep adding wood to it, my grandmother used to say when I was a child.

The same goes for great leaders; when they think they are on the top pedestal, they fantasize and dream. Theories could go off track, losing all glory, because this pedestal is now a minefield, and his next step could be his last one, as he plays with hope, the salvation of a doubtful future and faith itself.

9 What Is Beyond Death?

I have been bombarded with these questions by people almost daily, indicating thinking about death is everyone's pastime. Once in a while, I stopped at Weehawken Cemetery in New Jersey to see if my daughter's gravestone was still there. I don't know why cemeteries have locked gates, because people only go there to bury someone. They keep the motor running in their cars while there and at funeral homes. They go there, sign the presence book, give a glance at the cadaver, and run to catch the plane, but at weddings everyone arrives hours before the religious leader and stays until breakfast is served.

The problem is death doesn't show its face as a spirit naturally; only a few affirm having known a ghost. Everyone tries to take photos of them, but the cameras don't like doing this job.

Those who say they have seen ghosts or images of the ones now beyond our carnal world are mediums (those privileged with the capacity of vision of the fluid world), and they are the only witnesses of this phenomenon. Like everyone, I would like to enjoy the club.

The United States has an abundance of ghosts, giving Hollywood all the necessary material to keep the industry alive. The Award for Best Picture or the Oscar winners every year include a ghost story based on the invisible. It is being sought by the visible in the hope it goes beyond the limits. The movie *Ghost* hit the movies houses and also hit the TV screen, and everyone ran to buy the tape (the DVD came later).

There is suffering at funeral homes, where we hear the screams and see the tears of disconsolation, because the one gone from us doesn't animate the carnal body anymore, as he or she was known to us. Now it has become valueless and must be discarded immediately, because it's nature's law. (In most countries, it must be in the ground, incinerated in less than twenty-four hours, or frozen).

The polemic of us being left behind naturally to follow then in a very near future is not an option. The average person believes the dead are now at peace, away from the sickness. If this is the case, they are gone to "rest in peace." I don't agree, because I don't want to rest but be active in peace, because in the spiritual dimension, I bet life is even more colorful and thrilling than it is on earth. I am prepared for that, and I better be, because when you get into your eighties, the worries about the afterlife are accepted. Everyone sees a portal engulfed in a mist, but there is no way to peek into it, and no one dares either. That creates a feeling of emptiness and fear, and then all the tears go to the departed one, but time heals the pain, especially if there is a birth in the family, wedding, or any other such event around the corner.

Our thoughts never sleep, and that is good. It keeps worries at bay; it is as if all the mysteries have been resolved and the contact with the spiritual world is at hand. I believed it wouldn't be exciting anymore. It would be tedious or monotonous. It would make life worthless, and suicide clinics would be more popular than coffee shops and restaurants. No one would worry about Hawking's black holes or, worse yet, his godless universe.

Religious preachers' faith must be unconditional. That means they must believe without doubt or questions. They just have to accept it as being instructed by the leaders, according to the writing in the holy books. With this faith, there is no salvation and much less hope. I see those preaching as having a college degree in religion, and if someone wants to follow it, it is a great idea because that follower goes to bed and sleeps and is happy.

At my age, I became like an indirect guru. Maybe, it is my short beard, mustache, and white hair, giving that impression of already having one foot here and another in the great beyond. As I am walking along the streets, at restaurants, in bars or shopping centers, on the bus or cruise ships, in airplanes, or in a taxis, people will ask me, "Sir, sorry to ask questions, but it has been bothering me for a while. Could you explain to me what life is like beyond death?"

I have become so used to that question that it doesn't bother me anymore. Now it amazes me to know I am not alone in this matter.

Some souls argue because they are unhappy about questions that have no logical answers to please them. They are mysteries that set them on a dead-end road. The theories are good for kindergarten children.

I can recall back to when I first learned the alphabet. I was still sucking milk from a bottle. I had a pacifier hanging from my neck, and my light-golden curly hair was way down my shoulders. My teacher in high school told my mother she could not believe my questions and answers for my age. My mother used to say, "We are not created equal, or God wouldn't be God!"

Even after medicine declares someone dead, surely only about two hours later, according to the temperature, the body begins the natural process of decomposition, faster than any carnal animal. This is nature as unchangeable, not allowing the spirit back to it.

Because we are in an area where the Bible influences the Catholics and Evangelists, I read my wife's Catholic Bible with 1,200 pages not missing one sentence, and I did it reading one hour a day for two months after midnight, so I could discuss it with her, relatives, and friends who have read the first and last page, as churchgoers, but loved to tease me: "How did Jesus resuscitate Lazarus after four days of being dead?"

I always said, "I am not a preacher or a religious father" and to ask them. It was a logical answer, but they looked at me with disappointment, as if I were hiding something. As always, when it came to family, I was the loser, but to punish them in the last big dinner where the freeloaders all came in, I did not enter the kitchen. The dinner was then as ordinary as it could be.

The only thing guaranteed and confirmed by science and anyone with common sense is the aging or breakdown process of our bodies. The same goes for the vegetable world. Now, in the twenty-first century, the DNA has been computed already in the big bang, because it is as infallible as nature's law (God's law), programmed ahead, as is everything for our visit on earth through the fluid as human beings.

The interesting thing, I believe, is the big brains, the famous ones, are dedicated to reaching the unreachable cosmos. It did not move one inch in the direction of creating apparatuses or gadgets to discover where they belong in the spiritual world.

The telescope reflects light from millions of light-years away. It is just a dream that we will see it but just looking at it is a hobby, or pastime, while in the microscope, the microbes are still laughing at us, but how about the spiritual world where everyone goes and stays, including our close relatives, friends, and enemies? Stephen Hawking knows our time is getting shorter. He is ten years younger than I, saying lately that he wants to live more. The reason is his fan club pampers him. As a godless man, he is also a soulless man. When death comes in, he would then exist eternally, and those thoughts have him terrified, as time is closing down his curtain. He would rather be a mummy than a dead corpse.

Let's stop looking at the spiritual pedestal as if they are only statues of saints made of stone, marble, wood, or clay but as a pulsing spirit, dancing the tango as seen once, by a drunken man. For those who confirm it, I would be the first to vote for them to receive the Nobel Prize as a brilliant mind.

Every day, we step on a scale worrying about our bellies getting bigger or our dresses getting tight and not supporting our curvaceous shape anymore, but no one ever stepped on the digital scale and felt a spirit verifying his or her gravity.

Many of us say the spiritual world doesn't have the afflictions we have on earth, but that isn't true because when we arrive there our debts—moral lapses, laziness, arrogance, ignorance, bad faith, infidelity, crimes, abuses, and so on—will affect our spiritual life, because it is part of our conscience. Being rich and famous are earthly as material statuses, where there is another style of life as a third dimension, a society where values come with sentiments. Jesus told us to love each other as we love ourselves. He knew the spiritual world or he could not have accepted the cross.

But no one should be scared to go there, because there is no choice. It is our destination, a place for the ones who were discriminated against because of their status as poor, ugly, or not so smart, and those to whom doors were closed, there is a spirit with a perfect human body. He was fair to others here. His records or DNA would be classified for better positions but never on earth, as social status classifies the ones who had all the material greed and the ones who have nothing but their labor.

As is confirmed by the powerful Hubble telescope and other great apparatuses, available thanks to the continuous research, especially by NASA and the Russians. The knowledge of deep space is getting more sophisticated and helping humanity to learn and to give advantageous technology in research that brings us comfort. It makes things better at home.

But the ideas of other planets like our earth, as common sense tells us, is more than possible, but the reality is the immensity of infinity. There are even more universes, because our big bang was reality. Why wouldn't there be an infinite number of big bangs creating conditions for life like ours, as in the past? Why wouldn't it continue eternally, not as possible but real, as we see only our backyard, which is not reachable in our perishable, carnal body.

We can't discard that we have and are having contact indirectly with human beings from other planets with a technology above ours, as the infinite is just so vast that it could even be considered as nonsense. We believe only earth has intelligent life. So what good is life anyway, if it doesn't continue as an afterlife in spirit form in a continuum, as it began? All the existence, intelligent or not, carnal or spiritual, here or anywhere in the hot guts of hell or in the cool infinity of heaven, for us, as intelligent, rational beings, is as astonishing as the beginning. Having an eternal continuation as spirits in other dimensions goes beyond an afterlife, because we can dream about conquering eternity, which confirms our thoughts as reality, or I wouldn't be writing this book at my age. I didn't do it just to put words on papers. I am sitting at the computer with my feet swollen and my right hand hurting a lot because of the mouse. I could easily do something else, such as driving to the mountains or anywhere or taking another cruise. I feel this book is a communication, and that's all I can say, as you imagine. That is why intelligence makes a difference, and it can't be wasted with the end of the carnal body, because the Creator's investment is real.

To the ones who will alleviate themselves of it spiritually, I say good for them. It is a blessing, but I must do it as a mission and then take off, like in my Cessna plane. This time, I will trespass gravity into the glow of the infinite dark matter, taking my spirit beyond the negativity of earth to worlds only reachable fluidly, as fluidly as we came here as souls. The beginning is here in our big bang, as we race as a sperm to an ovum. It's not a joke, unless they, as "brilliant minds," have the theory of the month, like Stephen Hawking's quotes. Anyone can do it. I could write a good thousand, but who cares? He must love to be a monkey and run from the wheelchair. As he said, "We are just an advanced breed of monkeys on a minor planet of an average star." Here is my answer for him: "Mister Stephen William Hawking, the most brilliant mind and a genius with a 250 IQ, while you are a mummy nailed to a wheelchair, your mind is very busy bringing hell to mankind. You are as disrespectful as a man can be. You are inactive, as the effect of a cause, and this cause is your disregard of anyone and anything including earth, calling it a minor planet of a very average star. If you don't like it, just jump ship, killing your carnal body. You confessed that you want to live in this 'not so beautiful body of yours.' As you mention, you can see women but are not able to touch them as a spiritual punishment to an individual. Society sympathetically changes your diapers, feeds you like a baby, and takes care of what is left of your carnal body while wheeling you down a red carpet. What shocks me is how little you are giving to humankind. All you are taking is beyond my reason. While you are tenderly being carried, as a blow, 'believing you are a brilliant mind,' you just piss in everyone's face. You believe we are a breed of monkeys because you are one; otherwise, you would not talk like one. Your voice is being translated by a computer mechanically, because the technology created by humans, as an experiment with computers, let's you do your godless monkey talk. Earth doesn't have enough trees for you to reach the stars.

"If you really understood the universe, you would have found out love, charity, and a supreme power did orchestrate the illuminated dark matter. From it came the universes and you, as a special *breed of talking monkeys*, as in the successful movie series, *The Planet of the Apes,* as Charles Heston the great actor (2008) interpreted it."

Hawking should watch *El Cid* (an eleventh-century Spanish warrior, who defended Spain from a Muslim invasion), as also played by Heston, to know how primitive human beings still are but still ahead of monkeys. Even in the name of God, his punishment is those lacking love. Imagine earth being godless.

There would then be just one moral law: "One for one, and no one for anyone." Christ came and died for his principles, such as: "Love your enemies, and then there wouldn't be any wars"; "Love one another to receive heaven as a reward." They nailed him to a cross. He begged our

Creator to forgive the ones who crucified him because they knew not what they were doing.

Hawking has been nailed in his wheelchair for a half century, godless and proud of it. He doesn't know he is taking hope and putting anarchy in a world full of Hitlers, Bin Ladens, Saddam Husseins, Bashar Al-Assads, Gaddafis, and so on. The list is always getting longer. The latest one reminding us of his big face is Kim Jong-un, stepping like a pig on the lion's (the United States) tail with a lethal torch (atomic bombs) to be roasted, as in a pig roast if he lifts it toward the lion's face. It is not an arena with just two in the fight, as in the *pig versus the lion*, because it now involves millions of innocent people, as the population could be in a holocaust.

If Hawking was a real brilliant mind, he would be doing all he could do to save this minor planet populated by monkeys, but with his nonbrilliant mind, he plays the heartless blasphemer, as a monkey has no pity for mankind, like in the *Planet of the Apes*.

Physically, a monkey looks better than he does, because to be human with an ape face with the teeth touching the nose is not too graceful. I will invite a trained ape for dinner and then a monkey disguised as a human scientist.

Those immensities spot the infinite knowledge of galaxies floating in the endlessness of the glowing dark matter. It wasn't created for us after we were born here to melt a pile of sand. We make giant lenses from pieces of glass and digitally aim it to infinity believing we own the universe, but we are going there anytime as a surprise, because when the zygote enters the uterus, it gets a passport called the death of the carnal body, and our free ascent to the faraway worlds are guaranteed as spirits.

Proof of it occurs every second; death takes its toll, beginning now with the pope getting ready to depart, as did my thirty-three-year-old daughter and 251 teens and university students in a nightclub fire a few weeks ago in south Brazil with another eighty dying from the smoke inhalation. They calculate that a minimum of 240 million died in World War II.

This war alone was five years of absolute hell only sixty-eight years ago. I was twelve years old, and mercifully, at that time, due to my age, my worries were only about studying, playing, and swimming. I won many medals. Brazil was in the war, but its citizens were away from the killing fields, as was the United States, but our soldiers were not, as they guaranteed with the risk of their lives that our planet would be free from the evil. Evil comes disguised in many forms, challenging our evolution and spirituality.

We must continue to keep our eyes open for the ones infiltrating our society, giving us what we don't need. It is anarchy, making a mockery of our intelligence as if they are here to help. They act like they are super geniuses dotted with "brilliant minds." Everyone else is a jackass, but their nose stops at the stone wall called common sense, as reason put them where they belong. The rainbow always shines after a storm.

A phenomenon itself is the talk of science, especially Hawking, who spends every minute around the clock with his nose at the monitor, not doing research to help eliminate the microbes that eat our bodies and soul, but creating theories to "break God's chops," as the English said, while I was in London, savoring a kidney pie and drinking warm beer for the first and last time.

The theories about black holes that will, billions of years in the future, reach earth and eat the dust of our bones or about the sun in billions of years incinerating our blue planet. All the souls as spirits will be up there in one of those galaxies, where are all the past generations are, because they had hope, knowing there was life beyond death. That is the way to continue living. The pharaohs millennia ago prepared their entrance to the galaxies, and at that time, they didn't have telescopes or monitors to put their noses to so they could create stupid theories. The media has no better news to print. It has the priority of keeping those brilliant minds on their sand pedestals.

When I look at the pedestal of life, I see God, but Hawking sees a monkey. That's the reason he says it not as a theory but as a joke. Seriously, his vision of human beings is that of monkeys. If he had a normal physical body, he would have taken over earth as the anti-Christ people are afraid of. God respects our free will but will always curb down the big fish—aliens disguised as scientists, because earthlings have a shield of faith in Christ. Hope is the backup.

Too bad I can't come back in few years as spirit to straighten things up. It is God's will, presenting me a glorious 360 degrees road to navigate my Cessna beyond earth's gravity in the crystal-clear dark matter. A few years ago, above the cities of the great nation the United States, which received me into her arms and gave to me as to others the freedom to express my soul in the direction of love to all, they say "freedom to all." Just after I came to New York City, Brazilians fell into the dark ages of a merciless military tyranny that lasted twenty-one cruel years, ending in 1985. God kept me under the protection of the Statue of Liberty, so I could reach eighty years of research as its citizen, to then be able to write this book, finalizing my duties on earth. The States gave me the opportunity, but with the Creator's blessings.

The black holes seen by Stephen Hawking on his monitor are here on earth, right under his feet. Anyone who is a scientist or an imbecile can dig it up with a pick and shovel. It was confirmed by him, when he stated that even in his precarious condition, he wants to live. As he said, it is up there so far away from our homes that if it reaches us, it will take billions of years to get here; meanwhile, NASA will be ready to handle the situation.

I don't consider theories, maybe a nullity, because theories are not factors or pictures, but dreams, like flying with wax wings. They can be melted by the sun. Blasphemy is a mortal sin against the logic of a Creator. It ends in severe sufferings. We must wake up, because we do not create, but are created.

Above everything is the reason based in common sense, because faith without reason is not faith at all. It is above theories, as theories are based in dreams because they don't come true. It is difficult to distinguish it from reality, as it is a play on our consciousness while we sleep. It is a fantasy, something we know can't be reached while awake. Truly, I never wanted to listen to anyone's dreams or nightmares, as they are not reality; otherwise, I would have had a million lives in just one.

Our limit begins where the Creator originated us in the total or dark matter (Universe-Mother) because it involves all that is in existence. It is his home, and because of that, it's neutral and doesn't interfere with any compositions created as universes. It is the aquarium, and all the existences are in it. It is not hard to conceive of it.

The universe or universes are nothing more than little fishes in an infinite aquarium. The glowing dark matter always existed, even prior to the life of this endless ocean, before the past was the past. The present time is always the present time, because it is the future. The future is nothing, as the present is eternal or there would be a future.

Science never plays with the afterlife. Even the majority mention the Supreme Intelligence, as Einstein did, which I mentioned before. They wisely leave it to religion to play with it as they like. Religion is losing its faithful followers, because the illogical theories are in competition with faith, and the battle continues. They need to follow today's changing minds.

My dear readers, if those problems in our lives get more confusing, just think and remember nothing is given to us on a silver platter. We must earn everything, and in the end, we will get our rewards. Sometimes, it begins here, but it is guaranteed on the other side, because the universe is us; we are part of it.

One of Einstein's theories was discredited, and he admitted it, changing as evolution rolls. He had elasticity, because he was a "brilliant mind." To admit is to win for everyone's glory. Remember, after one genius is another genius, because time keeps ticking, as does evolution. God keeps creating. Lavoisier brought only scientific facts and not a package of theories. He proved that nothing is lost; it is transformed, and it gave us logic that contradicts the black holes. It is just a dream of the undreamed; it devours even light. It can be digested, but digesting doesn't get rid of what is already created, like an atomic explosion. The atoms will continue as irradiation and will be together one time in the future.

We have daily affirmations, especially from the United States, England, and Brazil, of people, including children, seeing and hearing "ghosts" or spirits, friendly ones and scary ones. Photos are presented, but that doesn't convince the ones not privileged but makes them curious enough to pick on this phenomenon. There are houses available today for rent, as "ghost houses." They risk their necks for confirmation of the afterlife. It's interesting, but religion and science are keeping their distance.

Brazil, like India, has always had great phenomena of healing by mediums (people capable of having spiritual contacts), and in the name of curiosity, I visited many of those real "Saint Thomases." The Catholic Church had to see with its eyes and not "with the eyes from someone's mouth." From those adventures came my book: *Dr. Fritz: The Phenomenon of the Millennium* (2002).

Now, here is my question: why do religion and science not get together on these spectacles and everything else as a family? This question is the same as: why are there wars? Any answers will be the effect because they will not reach the cause, at least at this present time, because moral and physical pain must reach the souls in one single blast and not slowly, as in World War II. It didn't officially hit all the hearts.

They do consider it a waste of time, as something as mindless as God itself. They ignorantly talk about an invisible ghost, acting like Saint Thomas. There was no talk. They have to see the footprint of the lion for it to be real. It must jump at you for its a dinner. It is just my right to express it this away. It is because the Creator would jump on the scene, and everyone after that will have to behave. There could also be the terror of facing the Almighty and then losing the material status of a fake supremacy, like the pedestal on the beach that means it will be washed away on the next tide, such as sickness and death.

I know this is more than my imagination and doesn't go into theories, because millions of people, including famous ones, agree. For this reason, I did not wish my infancy. No one worth the Nobel Prize is evil. Every night up to my teens, I saw spirits and ghosts, because I was awake prior to being asleep. They used to appear to me in different forms, guaranteeing I would run to my parents' bed and jump among them with my heart pulsing so hard it was almost coming out of my chest.

If it wasn't for my mother as an angel, I wouldn't have passed my adolescence and learned to not fear the fluid world, because earth is the souls' territory.

The visions are more frequent to children. They come as angels and to some as devils, making it a real hell on earth. If it continues to the final act, the poor soul can land in a mental institution. I had help from many adults, including Catholic priests, and the spiritual world left me in peace. It was a space for angels, and I am grateful to know we are eternal. Those who have the blessing of seeing and feeling the invisible world know that God exists, as do we, in a dimension not reflected in our retinas. The spirits are seen by our minds, and that's the reason in ambient light only one or a few will see them.

The "proof in the pudding" today is the Vatican with its "elite troop of exorcists." It means expulsing bad spirits from someone's body or mind and not blaming Satan and his associates.

This important job requires them to go anywhere. For speed, they have a fleet of small jets ready to go with specialist priests, who have a crucifix in the right hand and in the left a sprayer with holy water. Sometimes, this ritual can even take hours. When spiritual things are out of hand, just dial: VATICAN-SPIRITUAL-EMERGENCY, give the address, and they will be there as merciful angels, dominating in the name of God and Jesus, with the help of good spirits. Those allowed to come give lessons to us about a life beyond the Valley of Death and then are called back.

On April 30, 2004, the *Los Angeles Times* had an article on the first page: "The Vatican Top Exorcist Sends the Devil Packing, as Father Gabriel Amorth Does Battle with Satan as a Busy Man."

Verify on sites that "I am not playing with fire." Pope John Paul II, at the World Youth Day at the Vatican, was assaulted continually by an adolescent English girl. The pope himself, with all the help he could get, was losing the battle against Satan and his family, until the special group of exorcists landed and the bad spirits' fun ended. Now, in Rio de Janeiro (2013), there will be a World Youth Day. The image of Christ is 2,200 feet high, as the seventh world wonder. One looks at it and wonders if Satan will dare to show up.

Now, after all the evidence of two plus two equals four, the department of the Vatican as the number one in science affirms they should do better on their exorcism.

Create a sensor to detect spirits and then take good photos, and hire the genius or the "brilliant mind" Hawking to keep his eye on it—not with theories, but realities on the screen and in pictures. I believe he will find God, especially for his benefit—because his counting of black holes is not helping his life or ours on earth.

They dedicate their lives to running after particles, measuring them, seeking continually for smaller forms with a fancy Latin name. They are desperately running now after the composition of dark matter. Everything with the big bang is fine, but not to include the spiritual world is a serious mistake. It is just incredible that in his condition he is not waking up to seek the right medicine on earth, if not to appeal to the spiritual one, because there is no other solution. They just remain sitting on the nail.

It is not necessary to have graduated from the best universities to be a brilliant mind or to have logical instead of illogical theories, because the best institutions of teaching include religions; more than that is the Creator, even though he is personally not in front of us as a figure. He is felt by his creation, and we, as mortals, have proved in every second that we do not know about being here to see the next sunset, because even the airplane is not guaranteed in the ticket to take off and land, not killing all aboard.

As the dark matter is the residence of the Creator, we can one day come to know the composition of it, and that is good, because to research is to tune our intelligence and upgrade our IQ, but be careful, or you could touch his head, and then we must all ask for his blessings.

10 The Mind in Harmony Makes a Happy Life

Psychology is the study of the mind with the intention of helping with disturbances, such as depression, which is common, and also problems affecting the normal conditions that lead to us having a normal life, or better yet, aims to increase a person's well-being. This program would work best if it were done by angels and not by others who are on the same level as the patient. I have had many negative encounters with professionals related to friends in their care.

The ones dedicated to this vast field, being men or women, are working in a minefield; the next patient could be the one who steps on the mine, bringing chaos affecting not only him or her but others. The human mind is the most complex area. The brain is a mass wherein our soul sits, because this is the thinking zone of our being.

Even a new student asks this question, and it is so complex, I do reach a dead end. It is my right not to deny the Creator, as it is Hawking's right to deny a Creator and say the universe came from an inanity. Suddenly, in a big bang, it materialized, but Hollywood did not fall for this one, nor did the average people, including me.

In the study of the patient's mind and his or her behavior, psychology seeks to understand and explain how he or she thinks and feels and then to help, but as for the one with the concept impregnated in his mind that life comes from nowhere, it is better for all of us to have him confined to a high-class mental institution for "highly brilliant minds." From there, Hollywood could have a great theological imagination to guarantee continued entertainment.

But God made his corrections, because he is the first super genius to be betting by a microscope. There are no genius germs. He is not battling to win his glory to walk around enjoying life as far he could but sits on the divine "nail."

I was at a reunion in Manhattan many years before cell phones took over the planet and GPS was still a dream, when someone said that sooner or later, we would be able to read people's minds. I got up and said, "It will never be true; even the 'lie detectors' can never be perfect. Sometimes, I believe not even God could penetrate it, because of free will. Our mind is our private world, and if it is violated, we will just be hobos and life will then be meaningless, as anarchy takes over."

There are those involved in the practice of helping others when suicidal thoughts take over. The patient believes there is no one he or she can really trust or who wants to listen to his or her problems or he or she can't face his or her priest. I am the first one to point to the ones who graduated to help those closer to a breakdown.

I lived and live a life full of action, if I may say so, even now with my feet like balloons for having been sitting and writing, using the computer for up to twenty hours daily. I can't stop, but I believe that in my eighties, I am in great shape in body and mind, because I did not sit down to let problems take over. Life is not easy for anyone, rich or poor, brilliant or obfuscated minds.

As I promised my soul and spirit, this book will bring to the general reader what is not found in most other books; I will express realities, not to bring anarchy but consciousness to the ones off track, to bring goodness and peace into our stress-filled lives as humans. The next minute could be great, or it could bring a tragedy, as said many friends. We see our children leaving, but there is no guarantee they will ever be coming back. After thirty-three years, my daughter Carol (1996) never came back.

Unhappily, I can say that I have met many professionals in this fine line of psychological help to the ones in need of a consoler in their moments of stress. I have found the ones who embraced it, not for love, but the social status and financial rewards, while looking at the low cost of investment. But that also goes for any profession.

I noticed the existence of doubts that the human being is intelligent, that it is rational and different from the irrationals; they bring rebellion against the Creator in the free will. I have Stephen Hawking as the number-one example of it. He uses his status in science; by many he is considered the one after Einstein, who was and is godly man, because, as I affirm, he continues alive as a spirit. Every one of us will. When he left, he felt an energy involving his body, as it was his own fluid body being liberated from the carnal one.

Einstein had a rough life, but he didn't blame anyone for it. He was grateful to the Creator not to be confined to a chair. He did his best not to scare people; he even put his tongue out to the media, as if saying that in life, we should also laugh and not just cry and defame the illogicality of others. We should be thankful that he created everything for us.

Psychology, I feel, helps, but it is not the solution. It advises, because the mental problems are unique to each person's acceptance of the way our lives and existences are.

My niece was living in Washington, DC, in her teens; she used to say she wanted to become a psychologist, because she would know the patient's life secrets, while making good money and being respected in society. At the age of twenty-nine, she graduated, and her husband opened a fancy office for her, as she affirms that all she has to do to go home is to lock the door and leave to "break her young husband's chops."

To finalize this polemic about someone's mind, my next-door lady neighbor in her fifties, proudly announces she is a psychologist while she fights and screams night and day in her apartment. When she began bothering me, I told her she needed mental help and I was going to do it with my writing. She knows about my long letters to the not-so-good politicians and the results of them.

That night, I wrote and put under her door my letter to her about how to treat her marvelous cooking lady, the general teen cleaner, and the chauffer humanely, as they are human beings and slavery in Brazil was ended on May 13, 1888. I told her I was going to press charges. My experience in this field comes from common sense and love with spiritual understanding. Many of these professionals in this line, if they had scholarships from universities in heaven, maybe this needed planet would be a better place for our long vacation.

We were created to be happy, and an example is seen in the children. If we could be children eternally, then there wouldn't be wars and so on. Stress as the beginning of negativities would be gone. Earth would be a paradise. The worst is that the ones in this profession come from the same world we do, and they are at the same level of being affected as anyone else. It's difficult being in a carnal body with the same sentiments, good and bad, as part of our lessons as souls. Meanwhile, those people are claimed to be "brilliant minds," while the majority of the population gets up and does all they can to bring comfort and happiness by doing their jobs. They feel like just a number.

The only perfect brilliant mind is the One who came here, was obscure, and only in the last three years of his underprivileged life, lived knowing the way to let his word be engraved as short stories and a sermon that only lasted a few minutes. He kept it simple for easy understanding. He was surrounded by ignorant illiterates, but his wisdom was greater than all those finding the origin and the negatives of a universe without a soul or spirit and made by a heartless Creator. They are men without the meaning of what and how important love is. Just looking at infinity, he found what others, millennia later, with powerful telescopes could not see. He and our Father, the master Creator, wait for their creation to go back home. There is life in infinity, far from earth, and it only takes a glance to see it.

We are limited to seeing, feeling, and going beyond the carnal body, considered by the ones ignorant and insecure of what they are and must be, as being a zero. Their minds are not open; their thoughts confront a marvelousness, as rascals do when offered the paradise, and then they destroy it like vandals. When it comes time to pay the price for it, they look at the punishment and argue they are victims of the system.

Matter is our barrier on earth, as our carnal body, making hope our only way to journey and go beyond to another world. This hopefulness is being held by a thin string, but it is our *only* way out of the void of the tumulus in the emptiness of the cemetery, where the wind whispers, saying there is no one around from yesterday and no one will answer the call, because there is no one.

We are ordered to keep going, because soon, it will be our turn to come and leave our bodies, because our souls aren't there. Then there wouldn't be voices and crying, as there wouldn't be anyone there. It would be a "no-man's-land," as the silence takes over. Time keeps rolling, making the future today, because tomorrow is arriving.

Anthropology embraces the study of the behavior of human beings, trying to answer what we are psychologically more and more, but what if there was an "alien," or better yet, beings from another galaxy, if they were here because they are intellectually ahead of us? With their know-how, they could help us to make a better material world, but would they? We better change and help ourselves change for the better in every way, because we have here, everything at hand, as they have too.

How does the anthropologist get his or her knowledge? It is deep in their opinions; they will give an argument to those who do not agree with them! But "aliens" or even "angels" never will know where the Creator has his warehouse, assembly lines, or egg compressors. They will never know where his mines are in the dark matter or where he gets all the atoms and the kilowatts to keep his research lights bright. The eternity glistens continually without blackouts, but how about the illuminated dark matter as the whole?

The Cessna 182-RG has a maximum height limit of fifteen thousand feet, but the pilot without an oxygen mask is limited to twelve thousand. I went beyond the limits, up to almost seventeen thousand in the still of an ice-cold night, and the density of the air made it feel like I was in a gelatin-filled aquarium. I could stop the propeller and stay inert to keep gazing undisturbed, floating off earth's crust, fusing into the crystal-clear dark matter, seeking a meeting with our Creator and wishing someone would have been with me as an eyewitness, particularly Stephen Hawking, so he could feel as I did. He would then change his mind about there being a Creator, and the miracles would then begin.

This ecstasy I felt as eternal, and I just refused to wake up, but I felt like I was sinking silently and slowly, but continually, as the speed increased rapidly toward hell, giving the warning that heaven is up and not down. I noticed the instruments even lit in the dark, luminous red were obfuscated by the lack of oxygen. To recuperate, I had to push the stick forward, nose down, as the only defense against the diabolic whirl of the spin.

When my sense tuned up as I returned to a safe level, I saw the instruments clearly as in three dimensions and lit as by the sun, and I started a steady descent to the runway, which was as bright as a Christmas tree. After I greased the concrete (pilot talk for making a perfect landing), I floated to the next exit, and there I stayed for a half hour staring at the sky at sea level, wondering how blind mentally the average human being is, looking at such marvelousness with open eyes but not seeing it.

The ignorant gawk at the unreachable infinity without even noticing anything particular, but I looked at the sky encrusted with stars as a challenge, like scientists or religious leaders do when confronting the mysteries. With arrogance, they spread rumors and illogical events from their minds, causing panic and anarchy to people in need of consolation.

I gave the example to all that it is impossible in the universe to have an effect without a cause. Beginning with the dark matter as the origin and the expansion is not logical, because another big bang is getting ready to blast as it happens after a so many millions of years, because the construction can't stop, as space has no dimensions.

Many years ago, as a young man in Manhattan, New York City, I had a few classes about psychoanalysis. Freud (1939) was the father of it, and he had his successors, such as Jung, Reich, Lacon, and so on. After few classes or lessons, I had given to it all my intelligence in good will, but I felt was going in circles without an exit sign, and the monotony took over, making me sleep, especially after a busy day in my great restaurant near the Rockefeller Center, helping in the preparation of great gourmet dishes, plus all the problems a perfectionist could have as a restaurateur.

The reason I was going to psychoanalysis class was because a fine customer was challenging me with his awareness readings, and I needed some backup to take him off his pedestal. "Freud left practically everything to us to dedicate analyzing dreams, as an experience which explains and amplifies in debate involving religions, science, and culture," and the teacher was as eloquent, confirming positively that two and two equals four. I did get up, raising my hand, and was given the right to say something, but what I was saying was not what they wanted to hear. The experience I had in my early teens in a radio program, as the producer and speaker, gave me the eloquence needed to express what I thought about the dreamers' readers. I said a few gypsy lady card readers were doing a better job. I exposed to them my own dreams I had during my night while relaxing my conscience. They never gave me any logic on my life, and there wasn't a fancy name in science that would make me make a fool of myself, as I felt anyone could take a guess, just like the theories of science. Then, I politely opened the door and went out without waiting for the elevator. I never returned. I had spent some money, because few things in life are good and free, but from nature.

The majority of dreams are forgotten the second after we wake up, and they are gone forever from the memory bank. It only brings mental frustration from thoughts that are not real and without any relation to our daily life or even thoughts, because we did not make it up, as when we are awake. There are scenes that are never associated with anything logical to us. As we meet people and conversations or situations, it makes us wonder why these stupid dreams come as undesirable thoughts.

Our subconscious is just a nightmare and should be avoided if it is possible, but it is part of our mind. It is unreachable by any mortals. It is just amazing how we can go back and recall perfectly what was said in the past, proving we have a perfect data system to pop up as we wish. We can do it by ourselves. Dreams confirm that our mind never sleeps or relaxes, because it has an energy coming from our souls, and we can't control it.

People I love, like my mother and daughter Carol, both deceased, great trips, the thousands of people I met and meet in my daily life, remarkable events, my daydream of being a pilot that became a reality never but never came in as dreams or nightmares, while I slept or even dozed off.

In my opinion, Freud was mistaken about having dedicated profundity to a minefield, but without the mines to blow up in his face, because it was fun in the high society, always seeking fortune-tellers, card readers, or dream analyzers. It came to Freud while enjoying the best in times where traveling wasn't as hard as it is today, and communication was by pigeons.

Thinking and imagining what is in our minds is a daydream we have with open eyes; it is more realistic or more concrete and affirmative than what we do while we are in repose. The dreams Freud affirmed in his research are like seeking the dragon from the Loch Ness, the mysterious creature from the deep lake in Scotland—well-known but only in dreams.

When someone comes to me to describe his dreams or even nightmares from his sleep, I tell him or her not to be interested in anything like the fantasy from the mind, because it's illogical and it not worth a thought, or I would be by now in hell as a youngster or a card reader, analyzing dreams and nightmares, having fun as Freud did with the high-society ladies.

Some books we read stay in our minds eternally, and one of those I just gave a glance at while in the New York Public Library, at Forty-Second Street in Manhattan, stated the author believed that many psychological problems could be resolved by attorneys or at a great feast or at the police station. But I add, maybe they could be solved in a well-known voodoo meeting in Cuba or Costa Rica, while smoking fancy cigars and drinking lots of rum, tequila, or a good whiskey, and watching beautiful ladies in long, white dresses, swirling up to their necks.

11 Miracles, Phenomena—Real or Mysteries?

All the time, I see dramatic incidents or accidents deploring souls, and they grasp the spiritualism as their only hope, as if it is just a thin line that will give a tomorrow with fewer tears, and this help comes in the name of love. I do thank the Creator, but I accuse the ones without love for trying to cut off this line, especially idolaters, such as the brilliant mind, like Hitler and the Gestapo, and it only needs a closer group to start a global war, not the Korean's military group with that idiot-looking poppet, as a poppet was his genitor. As said Thomas Paine, an idiot lives with another idiot as his son to keep the anarchy alive while people suffer. But the French Revolution gave one sad example of trying to solve the problem with the evil guillotine.

Those faithless souls use their harshness of insensibility that is always on the tip of the tongue, causing chaos in the form of disconsolation. Not even the Creator can escape from them, but the effect, sooner rather than later, will be noticed by the impious, as physical and moral pain strikes. They have even been materially glorified. These groups are on the same spiritual level, unhappily as life is.

If I don't clear up these polemics, this book would be incomplete, as miracles and phenomena are fruits from the same tree. Those events, if I can express it in this manner, existed and will exist, because it is part of being human on earth and the effect of the free will.

Religion began with miracles; later came science as a phenomena, but the results are the same, as something that couldn't happen did happen, even against all odds materially. Not everything follows the track of scientific explanations.

Mysteries are the property of religions and science. When there is no logical explanation, they use these words to finalize their loss of face, when they are confronted by something beyond their reasoning, but to accept it, as being beyond normality, then comes the lack of spirituality. They all sit on a pedestal anchored on sand, because they stupidly are still loose in the name of the free will.

Acceptance should be normal when there is spiritual humbleness. It can be just saying thank you when something impossible becomes possible, as when someone is left to die at the hospital in the intensive care unit, corroded by cancer, and in a blink of the eye, he gets up and asks for lunch, or the big jet lifts up from La Guardia and its turbines suck in huge geese, and in one minute, it hits the Hudson River as thousands watch in horror but makes a perfect landing, as angels put it down gently on the surface of the river and all the hundreds of passengers just step on the wings without even wetting their feet. The pilot and copilot, when giving explanations of how the impossible did happen said, "We felt it was hopeless for those few minutes. We were powerless, going to hit the water with fatalities, but our only hope was to let God take over, and he did."

To top it off, the area is covered with thousands of boats and ferries, and a string of small planes like mine, going up- and downriver with home guests, to salute the Statue of Liberty at eye level.

I had my house in Weehawken, New Jersey, with views of the area over which I flew many times. As I run to view the big white jet just setting down like a swan, I told my grandchildren they had just witnessed a miracle. Someone behind us said it was a phenomenon, and I returned saying they are the same thing, because the spiritual world is sometimes allowed to interfere, but this citizen was a hardhead and said he didn't believe it. I smiled and said, "That's your problem, Mr. Hawking!"

As churches screamed, God put the plane on the water, and science affirmed and honored the captain, as the pilot did everything right, using all his skills and experience, while keeping calm. He touched the water in the right angle for a water landing and at the minimum airspeed while descending without power to keep the aircraft airborne to the last second before touching the water. He did the only perfect water landing with a non-water plane, without destroying the aircraft and saving all the passengers and crew.

This incident was an accident; it demonstrated the separation of faith and materialism, as the difference of the power of the free will and the presence of the Creator in our lives, but soon minds change, especially when the time to leave earth begins announcing itself. Aging demands its share as it has been doing since the creation.

As a private pilot, I did mostly night flights because of the beauty of infinity when darkness takes over our area. The Creator as always keeps the lights on. I felt many times a sudden aggravation appearing from nowhere, endangering the flight. I could feel the heat of hell and taste death, like when I felt a big shadow coming against me on a dark, rainy night. It was a tall building near my landing path, and the pilot's exclamation to God is automatic.

Once, at sunset, I was taking off from a small field near Albany, New York, and a flock of Canada geese crossed my path at one thousand feet. At least a half dozen of those over twenty-pound birds became minced meat and feathers. I had a propeller and not a jet engine, and blood blinded my vision. My only defense was to scream to God for mercy, as I landed back on the field. No one could believe I landed, except the Creator and me. I am grateful for all those experiences, because it brings us closer to life beyond the grave.

When a serious problem arises and science doesn't have an explanation and medicine has no cure, the ones in spirituality say it is a lack of faith, and they all use mysteries as a scapegoat, as the way out from the labyrinth without an exit, because research and prayer land in the same place.

When everything goes well, science says the equation was equalized; the second one says the dose was right, and finally religions claim the prayer was heard. Meanwhile, the Creator stays on lookout with an eye on everyone, because those three keep him busy.

Phenomenon is a very common word, because it applies to electricity, as the illiterate soccer player knows. In any sport, he makes an extra goal, like the one of healings, and in this field, science keeps its distance. But now, the Vatican, with the phenomenon of exorcism, keeps the bad spirits that wreak havoc with our behaviors away. It is an absolute evolution.

This enormous leap of the Vatican gives everyone a spiritual hope. It is evidence that those spirits are not Satan and his family, but ourselves beyond death and logically still around in the neighborhood, tickling our feet in the wee hours. For this reason, many of us cover them, even in the hot summer, when the air-conditioning is off.

I have speculated for years on this matter. I came to a conclusion in this spiritually controversial area, that matter on matter doesn't build or change without the power of energy, and spiritual energy is not seen by the naked eye. It is always present, here, there, or in infinity, because it is part of the dark matter as are the atoms, or whatever composition science calls it, in all forms—solid, gaseous, or liquid—intermingling continually, but respecting each one's bounds or there wouldn't be creations.

There is one challenge for science, and it will never be conquered because of nature's perfection. God himself knows the human mind, because it is private or we wouldn't be intelligent beings. We have our own world in God's universe, but it is our mind, as individuals, and it can't be penetrated.

The "lie detector" is based on the pulsation of the heart, but emotions have nothing to do with what's going on in the mind. Recently, in Brazil, there was a young man, nice-looking, with good manners; finally, police discovered he had emotionlessly killed over fifty human beings and would have still been doing it if he had not been arrested. Like him, the world is full of Satans disguised as angels.

Looking at people, one doesn't read their thoughts, and we can't distinguish their intelligence or what is on the mind. Intelligent, "brilliant minds" can hide their qualities. We often hear such remarks about someone that he or she "looks like a fool" but is a genius in mathematic or a successful entrepreneur.

Many times, salespeople would come to my restaurant, as they had an appointment with the owner, but looking at me as a young man, they used to say they wanted to talk with the big fish and not a waiter. I proudly used to tell them to get lost, as I was offended.

If it was today, I would be proud to be a young entrepreneur, but I am honored to be a senior. I am able to continue using my *Parker 51*, exploding words on to paper, doing more than creating fantastic gourmet dishes and making holes in clouds and running a radio program that opened my doors to journalism and writing.

An Indian back in history came to a city, and when he saw water coming from a faucet, he looked at it as a miracle, saying the water was coming from nowhere. The same idea came to my mind with the big bang, as it came from nowhere, as stated by Stephen Hawking. The Creator is not the provider; neither is the infinite dark matter, as the water to the Indian's mentality when analyzing a creation from nowhere. Sometimes, I wonder if the brilliant scientist got his ideas from the Indian fable or if he knows the truth.

But the Indian tale has a happy ending. When he made a hole in the wall, he discovered and followed the pipe and found the lake supplying the precious liquid. Going to his knees, he said, "Thank you, my great Lord, for my brothers, the forests, animals, birds, fishes, the sky full of stars, and the rain that brings water to quench our thirst because without you there would be no life."

The Indian story had a happy ending, because he had common sense and awareness that a spiritual power built our existence and it is around, fused into nature, camouflaged. As a good father, he keeps his eyes on the distance to control the behavior of his children. He could walk and appreciate nature and everything offered to us, because to get up and enjoy our moments here is a privilege.

This simple tale of the Indian teaches us about the concept of the big bang, now with an array of black holes, not as a decoration, but a hell from a mind sharp enough to see it. The result of a continuous staring into the infinite dark matter, not at fifteen thousand feet as I did, but in a plastic monitor's digital projection, seeking in it an illusionary dreamlike Creator. He wanted to show the planet he didn't find him, but frightening dark holes, as only an absentee God could defend us.

He is guaranteeing his status above everyone, but meanwhile, his life is one no one desires. It is easier for an uneducated Indian to recognize the perfection of our universe, than a not–Nobel Prize winner, researching with the Hubble telescope and mentally finding the Creator is not up there, but we all know he is everywhere, because he's above all his creation.

The vainglorious are usually backed up by a large group in society, as the ones who never pay attention are subject to the sacred law of cause and effect and of imagined happenings that everything is possible. Impossible would be our bodies coming from a zygote, making the impossible possible, as minerals also have life and because everything by science is possible. One who ponders, being intellectual, has his or her destiny ending at a grave, because to them, a Supreme Intelligence means they are just a number in the countless infinity. They feel like they are being discriminated against as a simple ordinary atom.

When I visited a small town, ensconced in the Andes in Ecuador, thousands of feet above sea level, I was in my forties and could go up and down in those valleys, and as I crossed a small agricultural farm, an Indian invited us (my wife was with me) for a tour of his manicured plantation. After a while, sipping their good tea and having cupcakes made from fresh corn and a soft cream cheese (umitas), I dared to ask him, a senior in his seventies, about the universe.

It was a sunny but cool afternoon, and the next city was the centenary Cuenca, about an hour's drive away, and the humble man lived away from universities, as he dedicated all his life to supplying us with what the good earth offers us. Looking at the sky where the ice met the dark matter and taking a deep breath, he said, "This egg from which the universe began had to be divine, because behind it is the Creator. Without an origin, the egg wouldn't exist, because to have eggs, first, I have to buy a chicken or any fowl. If I want to buy eggs, I know they came from chickens. If someone tells you there is no Creator, tell this person to come here and at night, without a telescope, to stare at the sky, and his mind will change seeing a clear dark matter, encrusted with stars and will catch a 'shooting star' as a bonus. Then he will realize the grandeur of our existence. If this soul, after facing this spectacular display, continues blaspheming, then something is mentally wrong with his mind for sure."

I never forget my conversations with the simple, humble people, without the grandeur of vanity. We hear words of wisdom, learned from the One who came here and stayed eternally, because his parables were more than short stories, but counsel to all of us seeking consolation and answers not found in religion or science.

This man was more than a scientist, religious leader, or politician, but all needed to be an eternal consoler, because what he was saying had more than counsel; it had the explanations necessary for all humanity to understand "why life is like this." From the beginning of our eternal voyage beyond the death of the carnal perishable body to our fluid dimensions; it is real, as real as our existence.

Thomas Paine, in his book *The Age of Reason*, said whether someone believes in Christ is irrelevant, because the concept of morality and love is so great that it fits anyone's psychological needs. It is working, so let's keep calling it Jesus Christ.

He gave us hope in any circumstance and not the menace of voracious black holes that swallow light eternally, while going over galaxies that engulf them. People are scared witless, because there is no defense or place to run to. Those black holes are a creation from the fantasy or dreams of a brilliant mind that spent a half century staring at a monitor, observing continually the dark matter that gave his retina black hallows, and now he passes them to us, because as a godless man, he has nothing to worry about, just the nightmare of an eternal death. He even now killed philosophy, while standing tall in a sand pedestal. He knows death will engulf his disfigured body back to ashes. As for his soul, I don't know, because I am not a clairvoyant anti-God and anti-Christ. A godless person has no heaven or hell to go to, but has to be in an eternal blackness, a black hole.

I have many Muslim friends. My daughter Carol (1996) married a fine Egyptian, and they loved flying in small planes. They became my favorite passengers.

One bright afternoon, the four of us flew from Caldwell, New Jersey, to Martha's Vineyard to have a great seafood lunch and see the beautiful scenery of the seashore. At four thousand feet, suddenly, we felt a jolt, and in seconds we were tossed up like lightning and then down. My friends yelled the word, "Jesus," and then everything went smoothly the rest of the day. Coming back, I asked why they had called Jesus. The answer was Jesus has the key to heaven, but while flying back to Caldwell, many times, I heard the short famous sentence: "Thank you, Allah, for being merciful!"

I will never forget that as we landed, they all embraced me and said I was more than a grandfather to them, but an angel our creator put in their life. It felt wonderful to know this world is full of good souls.

A few days later, the oldest came with his wife to my home in Weehawken, New Jersey, and gave me a golden medal from Isis, the Egyptian goddess of flying. Pilots in their country carry it, as Allah's backup.

Oil, as petroleum, was formed and transformed, according to science, over millions of years, and this process goes to everything, transforming all the elements but not destroying them, as time continually moves ahead and the same goes for our souls, which become spirits. If someone has another idea or theory, please don't tell me.

From the guts of the earth, we dig for the black liquid gold, but the theory from science as to its origin and creation is not very clear and never will be, because any composition here or there didn't just pop up from nowhere. It has been somewhere, in a distant time, but always rolling to the present.

Just look at the Hubble telescope. The immense explosions or big bang is creation in action. The composition of gazes have all the DNA necessary for life to flow as it cools or condenses to the purpose it is programmed for. In any big project, there is an architect behind it.

The ones not very interested in astronomy or cosmology many times know reasonably even more than those with their noses at the monitor or telescopes for over a half century, because intelligence or common sense is above theories.

As for my pen and papers, they were created somewhere as premeditated in time and space, like our bodies. Its composition was already in the zygote, including our minds and the intelligence of each one of us. No pills or injections are bought or ordered in any scientific laboratory, because our line ends where his begins.

Most of all, inventions and right ideas come from the general public and not only from those few qualified as "brilliant minds or geniuses," and this expression is the only thing on earth that irritates me.

Practically, I had the opportunity to meet few of those in this category, and we terminated the interview as I noticed the majority only spoke in one degree of knowledge, while they had no idea what was around their noses. My mother advised me to navigate life using all the 360 degrees; otherwise, I would have been lost. Later, after my sixty-fourth birthday, I needed it or I would not have found the runway, or better yet, any runway.

Atheists or the faithless go in just for curiosity as the doors are open at "God's houses." They see the ones being wheeled in tears begging for miracles and walking out, not being pushed or on crutches. But if it they could have cameras hiding on the private lives of those accepting the Supreme Intelligence, they would witness one in one hundred healings and even financially find the golden goose.

The Creator only hears and sees the ones begging away from the fanfare and sincerely asking without demanding unconditionally for aid in their miseries. If the need is sincere in his mind, looking at the Creator as the one and only planner of everything and believing everything has a reason to be as it is, then if the supplicant deserves it, he or she will get up, the four fingers on the left hand will be attached as normal, and the list goes on and on from the past to the present.

The ones who consider themselves arrogantly in the perishable body as "brilliant minds," disapproving absolutely of confronting the perfection of the Creator as a material from the material, have no illogical sense. They will begin paying the price materially, and the evidence comes in all forms and shapes.

As an avid reader, I followed Stephen Hawking, reading his books, whatever was published about him, and his speeches and thoughts. I have never appreciated anyone who puts his nose up as the wiser one on this or that fantasy or imagination. The great scientists are the ones giving logic and not dreams of black holes way up, up in infinity as a menace to our eternal existence. This dreamer is dreaming about having a normal life, but is defying nature as a mummified soul. The answers are in the spiritual world and better be.

When something is considered to have no solution by science or the medical field, far away from the multitudes, pedestals, and prayer chains, the only lamentations that reach the spiritual world are from the one in need at the right moment.

During my glorious eighty years, I have been learning from A to Z in a school called earth, where no two faces are identical. That's why we have as many religions as there are colors of the rainbow, such a variety of food, and so on. I saw hundreds of people dying, and two days later, we had a gourmet dinner, which not even people from their families noticed or talked about.

Someone said something I was sure, and it was interesting: those on their deathbeds, suddenly got well. I noticed these were the miracles. Could it have been the spirit of death appearing to me? Now I wonder if it was real or not.

Even now, while I am rewriting this book, my right hand has become like a ball because of the mouse, and my feet up to the knees have become like balloons. My medical doctor told me to forget about being at the computer for few days and then just one hour daily. I paid the bill, as in any professional work, and went right back to the keyboard. At night,

I say, "Lord, I have to finish this job before I can travel beyond earth as a spirit, but is there a reason for our existence here in a carnal body, not as an irrational, that wakes up, eats, fights for survival, has sex for procreation, and sleeps because it is dark and awakens because the sun is up, and dies being eaten by each other, roasted on a grill, or in a pot?"

We as rationals go beyond the limits of death, because we can think and dream of an afterlife. We have someone to hang on to, and it makes us an eternal spirit. We sense everything as simple or not simple, like the gourmet dinner I am having tonight. I am a sophisticated, intelligent being, able not only to look at infinity, but to seek my Creator, not visualizing, but feeling his energy. Whatever happens, I will always exist to glorify him.

The next day, in the morning, or even getting up hours later for the bathroom, I notice the swelling is gone and my right hand is as normal as it can be. I can remember clearly my life as far back as when I was three years old, but with all the details. I believe many of us can do that, because no one has that ability exclusively or there would be a continuity of evolution. Behind the "brilliant mind" are truckloads of it.

Many times, wise remarks get lost in the wind as a joke or a bad remark. One thing I never did was to try to make fun of someone, because I don't appreciate it being done to me. I don't do it to others, because negative retaliation is the name of the game. When we need help for any reason and materially it becomes impossible, then we are lost, trying to call the emergency numbers, such as 911 in the United States or *190* in Brazil. For whatever reason, we continue in pain, but I, like many others, also know to reach for the Creator and his number is on everyone's conscience: *God*.

I thought it was yesterday, but time doesn't stop. Forty years ago, my parents' neighbor, a fine man, openly said God didn't exist because our life was too hard and death was the end. For all his studies and hard work as an honest person, there were no results financially. As an electrical engineer, he tried for years to improve a good small electrical transformer but in vain. As he cried on my shoulder, I told him to go outside alone after midnight and stare at the illuminated dark matter encrusted with stars to plead for help. He could continue fighting the Almighty, and it would guarantee him a miserable life, including his wife and children.

The next day, he called me for breakfast. He had followed my advice, having nothing more to appeal to, and felt something different. It was like he was involved in a kind of energy, and tears came cascading down his cheeks. He ran inside the house feeling like he was being followed.

The following day, I went back to New York, and weeks later, my mother sent a letter happily saying the neighbor had finally created his dream. He had given her three small transformers for her TVs and refrigerator, and they worked perfectly.

I never believed in luck or coincidence, but many of us sit down and wait for luck to strike on our dreamland. Meanwhile, time keeps rolling as we age in the wheelchair, expecting miracles or phenomena to drop from above, and die sitting on it.

Many scientists in the medical field, after years of frustration in research, end up at a dead end. They finally appeal to the spiritual world, as doctors lately. *Time* magazine in 2000 had a special issue on angels and surgeons, as more than half of the staff in every country hold hands, including the patient's if it possible, asking the angels to intervene for a successful ending to the surgery and then recovery; after this article, I always ask surgeons about it, and they all confirm it is the truth. The next day, I went to several newspaper stands to get more issues, but it had a record sale.

Also, in March 1994, a *LIFE* magazine cover story caught my attention. It was about the power of prayer and the American people talking to God. Most of us are seeking what we do not understand, and it is a good signal of progress toward better lives. Only one country has inscribed its money, including its coins, *In God We Trust*. It is the United States of America. That is a sure way of demonstrating faith in a Creator and life after death.

Americans pronounce the words *Lord* and *God* more than all the countries together, and it guarantees for sure them being the leader of all the nations and keeping the fragile dove of peace alive.

I was so impressed by the article that I memorized one of those prayers, as I called it absolute hope and faith. Here, I will repeat it for those atheists and faithless souls on the way to learning that life is more than suffering and dying: *"Hear my prayers, O Lord, and let my cry come unto thee! Out of the depths of life the song of the psalmist reaches toward heaven like pleading hands. In burning words it says what human beings in all lands and ages cried to say to God, 'Help me. Heal me. Love me. Inspire me. Save me from my enemies.'"*

What amazes me is how interesting life is for Americans, including President Obama, who was recently honored. The number-one no-God man, Stephen Hawking, is now passing his half-century mark of going around blasting to the four winds a godless existence, taking our faith and hope in a life after death. This is my opinion. You also have the rights to yours and to express your feelings without deeply hurting others. Our views are logical, and they have science as a backup; there is no effect without a cause, and Hawking is sitting on his own "nail," because his mummified condition is his godless theories. Most ideas many times are fantasies or dreams that will never come true. As I spoke to hundreds of passengers on the cruise, the majority believed Hawking deserves pity, because of his condition, but remember the old saying: "The tongue is the whip of the soul."

If Hawking cannot see the beauty, for instance, as the incredible sight that surrounds our existence and given the opportunity billions of us did not have to reach the general public, then his reward is to be mummified and then have the opportunity to change his mind as a hobo, but the Creator can wait. (See "The Nail and God's Dog!".)

We should think of all the electronics, such as computers, cell phones, and the almost endless line of whatever was considered impossible that is now possible, not due to the "brilliant minds," but as a gift from our Creator, because a chip or pen drive with a few grams of metal holds wonders hiding in it, as do our brain cells. How is it possible if it is not a gift from the Creator? Anyone with a clue please use my e-mail to let me know. If it is just theories, I will delete it, because theories are not solutions.

Open *sites*, and try to understand the composition of a *chip* and how it works! It will be a big frustration and surprise. After going through pages of research on electronics, you will get to a point of no understanding, as nothing is logical for us as humans materially. None of us are geniuses, and much less "brilliant minds." We especially do not understand our birth or the beginning of our world as intelligent human beings; otherwise, if it happened to you and me, there would be nothing, because our existence is ourselves. Like everything else, we don't know it is absent from our minds, because a soul is not a spirit yet.

If the death of our material life in our carnal body is the end, then it is the end of the universe the moment we close our eyes not to eternity, because there is then an eternity, as affirmed by Stephen Hawking, paralyzed as his end is getting closer. He values the carnal body as a world even in his condition; he said spirits don't exist, nor does God, because he never saw one.

Neither perhaps do you, Galileo, Einstein, or even the great names in discoveries and research from the past to the present time, or even in the eternal future, less the one who became the number one in defaming the Creator. That's the reason for his suffering. He lives his life as if crucified; the cross can be any deformity, like not being picked for a Nobel Prize, as he believes he is above all the "brilliant minds." His prize here on earth is to sit emotionless, forbidden even to die now, demonstrating the ones ignoring the spiritual dimensions have a "nail to choose from," as idiots, self-defeating. That distinguishes them from fools, but they have the intelligence to defeat the ones not prepared to fall into the hoax, as a game.

Society is always seeking or looking to the geniuses and "brilliant minds" to keep Nobel Prizes on the pedestal, and the illusionaries seek a solution for the unanswered questions, just like Hollywood with the Oscar and the motion pictures awards. If wasn't for fantasy or dreams, our passage on earth from a zygote to dust would be as colorless as the irrationals.

Flour, water, yeast, the oven and its heat, and everything else for the bakery to exist—we have no idea where it all came from, but the baker we all know came as a microscopic zygote, like the irrationals.

Imagine the grandeur of God's potential as a Creator. Now there are those who do not realize how small they are, even as "geniuses" or "brilliant minds." They create nothing but put together what was created. We are less than a microgram (0.00001) of a gram in the universe, but we all are accountable, like the dark matter, as compositions of immeasurably small parts. All parts are important; without them, we wouldn't exist.

Now science is going after it, but as I knew as a child, with common sense. I will tell any scientist including those from the Vatican with its millions of gadgets bought by tithes. I have sought God up in the infinity of the galaxies, but to get a powerful microscope from NASA, like they made for telescopes, because the Creator is right here, very busy creating us as human beings with souls and in his image. Otherwise, we would have feathers, rough skin, or scales. We have large brains to give him big headaches.

We are not even the owners of our bodies, the only world we feel and enjoy. It is the main reason for us to behave and give thanks to the Creator. A sad example of it is Stephen Hawking, who is now afraid of dying, after spending most of his IQ trying to convince the general public of a nonsense universe. We are just an advanced breed of monkeys on a minor planet of an average star. It shows how bitter he is about his life, motionless, as he tries so hard to impress the public. His words gets tangled with his brains, spilling words of bitterness, pulling everyone into his ordeal: "My goal is simple; it is a complete understanding of the universe—why it is as it is, why it exists at all."

This statement from him is proof of his mental illness, due to his body condition, because the formation of words describes the intellectuality of the person, as a painter or designer does his artwork to impress the public. Hawking is far from impressing me, as it is a joke to the ones with normal minds who pay attention to him. But what beats my sense is that the media pays attention to someone offering kindergarten theories. My very intellectual mathematician young granddaughter just tells me:

"Grandpa, he is just a deficient caged in his body, saying nonsense, trying to be a god, and if it wasn't for his total disability, we all would be in trouble."

God must use us for his messages, which come in every form, to demonstrate that our fragile world is not fragile at all, because it is just a transition, a school of learning preparing his creation for a spiritual world. Thanks to religions, the majority of the population holds to it in a form of faith, which gives hope, as I learned when my dear daughter was mowed down by a twenty-two-wheeler by the *Times*, on February 5, 1996, in Times Square.

Thousands of people, including employees of the Marriott Marquis Hotel, came to cry with us, saying she just came here for a short while as an angel (she was thirty-three years old). No one from science as a godless mind would dare say one word, or I would put them where they deserved to be. They did not come to say she just returned as a pile of atoms.

The evolution of the species includes us, physically and cerebrally. It is affirmed by science through research up to now. We came first as primates, like chimpanzees or, better yet, gorillas, and I agree, because I wasn't there. Theories many times come from children in kindergarten. Stephen Hawking's theories of a godless existence, as he came from nothing, is the reason, and no one would change my realities. He is paying the price, or God would then be unjust.

According to Darwin those "gorillas" after millennia became the species *Homo sapiens* (you and I), and bingo, there was another scientist put on a pedestal of sand. I hope this book doesn't put me alongside those geniuses, because the planet now has enough saints and scientists by the grace of the Vatican. Based on Darwin, Hawking considers himself a brilliant genius monkey; he should have kept his mouth shut, because when I demand his DNA, he will be put in a cage at the zoo, and I will be the first to throw him his favorite dinner: bananas.

Almost everyone knows very well that only the Creator holds the cards of wisdom. Whether we are *humans* or *Homo sapiens* makes no difference, because we are what we are. It's final because there is no choice. Everyone is happy as it is, because no one ever broke a mirror. He qualified himself as ugly and that includes Hawking, as he is proud of staring indirectly at the cameras as a celebrity. His anti-God propaganda called the attention of 20 percent of the population of *Homo sapiens,* as his fan club, except William Moreira. I have had one eye on him for the last half century and the other on our Creator. It kept me alive and healthy despite all the confrontations earth presents to us in our daily life.

As we become better souls, we become able to understand the monkeys disguised as *Homo sapiens.* The universe came from a big bang, and the evolution of the species took over; that is fine, but there exists a separation of rational and irrational. Why did *Homo sapiens* not divide into different intelligent categories? I want Darwin to answer this one, but Hawking has a big brain; he does not offer an answer to this question, because monkeys have instincts and not intelligence. God is merciful, unless they are the ones contracted by Hollywood for the movie *The*

Planet of the Apes (they forget to mention it was originated by a book of Stephen Hawking's: *We Are from a Breed of Monkeys*).

The Creator doesn't need any of his creation to judge or qualify his actions, because we are qualified as created. Each one of us is to receive what we deserve, such as walking or staying frozen alive in a wheelchair or a bed or being both mute and deaf, black or white, disabled or ugly, man or woman, intelligent or stupid, luck or unlucky, and behind all those qualifications is a reason or better yet, an effect from a cause.

When we are children, we don't behave in small things, but they are not small things, because it will reflect in our adulthood, our parents, and us. Later, we will put the disoriented ones, if I can express it this way, sitting down in their little chair facing the wall for a half hour to "meditate" so as to be a well-behaved child. Beating up a child belongs to the dark ages, and the retaliation later would be fatal.

Everyone knows Brazil wins every year the Nobel Prize in miracles and phenomena and the United States does in ghosts, which are more popular than the ball games on Sunday afternoon. It is all in the name of God, but most of the religions have empty churches, while the beaches are full of sinners making no room for the seagulls, but melanomas and fatal cancers.

In Brazil, everyone says they are Catholic, but inside the church seats are dusty and the only faces are from the statues by Michelangelo. At least the Muslims are holding better to their faith, and at Mecca, the annual show of faith makes Allah happier to know his creation are learning to be grateful and to love him unconditionally.

Rio de Janeiro (sixteen million souls) is now considered one of the most beautiful cities on the planet, because of its natural surroundings, including the ninety-foot-tall statue of Christ on top of a 2,200-foot rock in the middle of the town. The Sugar Loaf rock is 1,299 feet with its cable cars for sixty-five people, and the curvaceous black and white stone sidewalks on the beaches makes a perfect atmosphere for miracles, where the population look at the Christ every moment, begging with their eyes for his help in a country with so much and so little to give its citizens. The sidewalks have more potholes than the moon, and the few public bathrooms have no paper, including at police stations, but the subways are perfect, because there are none.

Its scenery includes great old houses and buildings by Oscar Niemeyer, an architect called the genius of the modern world. He gave us the United Nations building in New York City, with its face all in glass and with its curvaceous lines. He filled the world with beauty, and Brasilia, the capital of Brazil, is a show because of his 104 years of labor. In December 5, 2002, at his last breath, he had the pencil and paper drawing curves, reminding us of heaven. I love curves; see my designs on the last pages, especially for the ladies.

He will stay in history, not for black holes, destroying the universe and qualifying us as a breed of monkeys. For this reason, he walked to his last day, gracefully drawing churches to glorify life on earth, as a short and glorious vacation of the spirit.

Miracles or phenomena happen continually in Brazil, because people there accept the wrongness as physical, moral, or financial with a deep breath, saying it's God will. Somewhere in the country, people start getting up from wheelchairs and the line stays as long as miles. Everyone hushes for the blessing, not seeing Fatima or Lourdes, where people are rolled in and out, because to be healed required more than being at the place or saying they have solid faith. Once again, just go to amazon.com for *Dr. Fritz: The Phenomenon of the Millennium*, and it's not a black hole but a sacred sign.

There exist only two monuments recognized officially on our planet for their beauty and more than that, the inspiration for anyone's life while here. There is the greatness of living beyond reason, because life is beautiful in any circumstance.

At sea level at New York City harbor, visible by millions daily is the Statue of Liberty, standing as tall as the Christ in Rio de Janeiro, with her beauty and arm held above her head bearing the torch of freedom, representing all of us as free brothers and sisters under one God.

In Rio de Janeiro, above the clouds, at 2,200 feet, there is the image of Christ, but he is not crucified. With his serene face, he is visible by air, sea, or in any directions up to fifty miles, because he was guaranteeing peace and love to the world. He backs up the Statue of Liberty.

Everything unknown to human beings is an unthinkable, and for those explanations, miracles or phenomena come to hand, covering the wall ahead without tunnels going through it, or the fresh tradition as "we are still researching" and the millennia will give better cellular phones, medicine, trips to Mars for vacation, airplanes going here and there in minutes. As time rolls by, the eternal sky has no limits, and death is here to stay, because evolution is forever, as nothing is lost but transformed to form again and again as the raindrops come down and then go up as vapor, forming drops again. Our spirits do as well because we eternally are the sperm and egg as abundant as the glowing dark matter. From it comes life, and we are never alone, unless you are from a breed of monkeys, as is Hawking's family.

There are those not interested in knowing, but as evolution rolls, so does time. Those sitting on the wagon, saying "I don't know, and I don't care," in time, they will wake up, because evolution is part of the future of all of us. There is no hurry, because as time is the present, it guarantees our eternity.

I consider reality to give my examples of miracles or phenomena, which did happen to me in my glorious eighty years, but you all have the right to doubt it, as I hesitate about yours. To see is to believe, and the expression in the Catholic Bible demonstrates, we are always distrustful of thoughts: "I am like Saint Thomas; I have to see it to believe it, as Christ is alive."

According to the Vatican, the spirit of Lazarus went back into his body, and months later, he died for good when he was assassinated because of jealousy over the reincarnation of Jesus, and this sensation makes Jesus as the Christ the only one ever to come back in his carnal body after three days and then go to the place we all will go as spirits. After this statement, no one has or had the grace of taking his carnal body with the spirit.

The vision of Jesus alive is interpreted by *The Science of Spiritism* by Allen Kardec (France). Jesus appeared to his followers only as an apparition or spirit, demonstrating the end of our carnal body. It is just the beginning of the spiritual one, as death doesn't exist. God has free will, but some of his laws are immutable to all, such as to be created from a sperm and egg as body and soul, live as a human being about a century, and then die, as a spirit belonging to eternity. If we came from a breed of monkeys like Hawking says, I would be able to challenge his monkey quote and theories, because I would be at his monkey level. I love bananas, but as a complement to my gourmet dinner. I was offered a monkey dinner, decorated with caramelized banana puree, in India. I refused it, thinking it could be one of Hawking's children.

Today, technology senses the energy in the brain (the residence of the spirit, which commands the nervous system all over the body) is registered graphically in a monitor, and as a little ball follows a line with up and down signs. When the energy disappears and the line is straight, then we are gone for good.

Hollywood shows it often in the movies, as in soap operas, and it proves the person's spirit is out of the body for sure, but I have a better idea. It is guaranteed that two and two are four, as to wait hours until the odor reaches our nose. After, there is no return anymore, not even being bathed in holy water, because this law is irreversible for a king or a beggar.

Around 1976, I was in my early forties. I had a very serious incident when cutting steaks for my famous restaurant. I say that because it averaged more than two thousand diners a night. While I was on the meat-cutting machine, my four fingers on the left hand became minced meat. It went all over the walls.

My restaurant was a fish and steakhouse, located at the corner of 156 Street and Biscayne Boulevard in North Miami Beach, as a number-one place for the best food and price in the country: *Bill's Beef House*, as I guaranteed any food was gourmet. All my 160 employees had to be the best, as I proudly ran it like a Swiss clock, having a sword in my right hand and my heart in my left until I paid the price, as does everyone else.

I was the only one, beside a professional butcher, who had lost two fingers in another job, allowed to touch this necessary evil. This job is like being in hell, because the blade has big teeth and runs at high speed, and everyone in the business knows if you do not lose fingers or a hand today, maybe tomorrow will be the day. Pilots say the same, as if following the *checklist*, tomorrow might be your day of not putting down the gears (wheels) even in airlines. Twice, if the control tower did not scream in the last minute, "Wheels down!" I would have become a fatal statistic.

As it guarantees for everyone that life is not just to be born, take advantage, and then die but is a constant challenge in any field, even with good health and financial well-being. I could afford having many employees risking their hands for good pay, but I was proud to cut steaks in thirty seconds, all with the same weight as also for chicken and fish, but this glory can cause pain with blood. Now, it comes to me that it is a miracle more than thirty-seven years later that I am putting down in maybe my last book, as I was heard all over South Florida: "Oh God! No, have mercy on my soul!"

Seeing the fragmented fingers flying around like dust being tossed into the wind, I felt I was surrounded by angels and demons, all fighting to have God's glory. I was the dummy lamb. The waitress was in line crying and asking God why he did it to me, because I had beautiful hands and was the best-looking young boss he had ever created. When hell breaks loose, we all become saints.

As I felt the excruciating pain, I looked against the light; my only eyewitness was my hanging little finger, the only survivor of the irreparable tragedy. I put my soul at the mercy of the Creator. The four North Miami Beach police officers having dinner with their families were busy sweeping the floor for any trace of fingers, including the scrap-collecting box inside the butcher machine.

Minutes later, at North Miami Hospital, the first thing I received was the blessed shot of morphine for the pain to quiet my mouth, as the head surgeon said: "My dear young man, who do you think we are, God? You are a bright entrepreneur of a successful place, and you know we can do nothing with a small pile of fragments of flesh and bones. Having the floor scraped makes things worse. Only God can do miracles!"

Then, as he embraced me, I whispered into his ear, "Just do your best. Wrap it up, and I will be praying all night long, because I have a lot to do for our Creator as karma, destiny, and a mission. He wouldn't hold it because of four cut fingers. Please do it for his glory and yours and mine."

The next day at home, as I woke up after one hour of sleep, I felt like my fingers were moving painfully under a heavy amount of wrappings. When I ran to the hospital, this blessed surgeon was there, and the hand I have now is proof of the spiritual world, and after all those years, only almost unknown scars are visible. We can stay or not in the wheelchair; it is our choice in our free will and not his, the Creator.

"I never had seen a miracle before, but you pleaded for it. Look at your hand. It is perfect, but full of black stitches. I will take them out in a few days," said the doctor with his eyes wet as he bent his head.

I left the hospital practically without pain but crying quietly. Not only did I cry at that moment, but I do it daily every time I use and look at my left hand, as I see it and touch it. Now, at the computer, I use both hands and all the fingers, without looking at the board.

The meek will be heard in their prayers, said Christ, and I say that the blasphemers will die sitting on their nails in the wheelchairs if they do not wake up to the Creator; for those on the way to the grave, it will be a harsh one.

My test from our Creator is right at my nose. (I doubt after all those years and without the computer, the record would exist.) The little finger is very delicate, as from when I was ten years old, I could do anything normally, and if it barely touches I feel some pain, as an eternal reminder of the compassion from our Creator. I hope it will be eternally.

About one year later, I was doing lots of lifting in the kitchen, not waiting for the staff to arrive. When I lifted a heavy case of oysters with the miracle hand, I felt a terrible pain in a disc of my spine. The third disc moved just enough to have me crying for almost two years. It thought I ought to have more employees and use my pain time to write and read to upgrade my knowledge, using my capabilities with my pen, rather than physical strength. I directed the business as a commander and not the commanded.

My experience with Muslims, especially after my daughter Carol married a fine Egyptian man, whom she met at the Cairo Marriott Hotel while she was on vacation, to Cairo, is God is a deep part of their lives. Galil was a very charismatic person, and we met families from many different countries, making our home a social club, reminding us of how great life is. They were very happy from the first moment they met, and Galil ended up coming to New York. Their resulting marriage was a touch of different cultures blending as a great history. Galil prayed to Allah a few times a day, grateful to have met a perfect young lady. They planned on having a few children.

Eight months later, everything changed in a second. It was just like millions of tragedies that occur every year. Everyone is surprised, because there is no warning. It strikes, telling us we don't belong as a carnal being on earth. Carol was killed on the cold morning of February 5, 1996, at the age of thirty-three, on her way to the Hotel Marriott Marquis in Times Square. Galil cried for weeks, but holding with God and Jesus, we know that paradise is not here, but beyond death. He knew his wife was in a better place, and we will all meet there sooner or later.

Hope of an afterlife is all we need in those moments, and the ones going around blasting the existence of this hope don't deserve to be honored as "brilliant minds," especially based on theories and not facts. It's unfair not to me, but to all of us, especially to the one who qualifies in the open media. He is from a breed of monkeys but knows answers are not rational to humans like you and me. How did everything begin, and why? It puts him at his monkey-God level.

I qualified those as anti-Christs, as no one in their mind is worth a penny, because materialism is a dead end for life beyond the grave. Our lives as human beings are on the same level as the irrationals. What I am trying to understand, my brothers and sisters, as we are a great family on earth, is how a big percent glorify them. I call them impious. The Brazilian government is impious, and the ones on high salaries busting the country are more than ever close to living a first-world life, while 70 percent of the population is earning less than three hundred dollars a month. The cost of living is the highest on the planet, as are the public hospitals with no doctors or medicine and all the gadgets are broken-down. The lines outside stretch for many blocks, and every five minutes, one sick citizen in those lines just dies.

The school system is worse than it is in the poor African countries, because the teachers all left to get a better salary elsewhere, and meanwhile, President Dilma is visiting one or two countries weekly promoting a fake country. All the fresh and salt water is polluted as the sewer goes directly into nature, while there is no restroom for the walking population. Big stores, coffeehouses, and so on are not obligated to have what is necessity for our bodies to meet nature's call.

It is beyond my comprehension and reason how the population are so timid in not demanding their rights, as the French people did. The result was that thousands and thousands were beheaded on the guillotine, and meanwhile, an anti-God man is glorified today on the international media. He is saving our planet by taking away what we all need as a balm for our souls, called hope.

Tourists come from the first-world countries, as educated populations. It doesn't matter how Brazil promotes the beauty of the country and how the Brazilians are friendly, the country's tourist industry ranks the worst in the world, reflected in the lack of lines at the customs and the limited number of arrivals at the airports like a poor, small country.

Sometimes, it comes into my thoughts that if Hawking had faced his heartbreaking illness as it hit the heart and soul without bitterness as do others spiritually, his ordeal of sitting down mummified for a half century could have had light, but as he continually rails against not only the Creator but to us as his spiritual family as under one God. He became an anti-God, as a revenge. He is bringing darkness to our planet, wishing at his funeral there will come a black hole, gulping all of us with his fragmented cadaver into the dark guts of his imagination where no light can penetrate.

You have your right to also sit down and write explosive words on paper and not hiding behind the stage criticizing illogically, because Judas Iscariot was more than enough as a simple, loveless man, and love is a synonym of eternity as spirit or life would be a Hitler's hell. Christ stayed continually on history. I am not a religious leader, but a fair person seeing the world as a gift going beyond the unknowable as a spirit, as life grew from a microscopic zygote into a thinking, able human being. We are here as a result of evolution, and it goes behind the termination of the body, or I would not be writing this book, but fighting for a piece of meat like a hyena.

I write or type my books, not putting words on paper, but exploding them on it, because what comes from the mind is fast or would be lost. We are all different or our existence would be a tedious one.

We decided our son-in-law, Galil, should go back to his large family in Egypt—his five sisters and one brother are all very well-educated people—for his benefit and because his faith and ours on a pedestal were resting on rocks. God is in our hearts. The pain from a beloved, beautiful young wife had his and our hearts fragmented. Sentiment, many times, takes years to heal. When we said good-bye at the airport, he cried like a baby but was glorifying the Creator for our lives as souls and spirits and mentioned many times that God has his reasons or he would not be God. The universe is too perfect, and he knows the Creator will pardon Hawking, because his physical condition did not make a dent in his mind, but nothing can come from a zero point in spirituality.

Well, merciless moral pain is alleviated as time rolls continually ahead into the future, but let us keep the memory alive as a balm that soon we all will meet again.

All the time I was healing, which was considered materially as impossible but happened, my wife used to tell me I only asked for mercy after carrying my cross for a while as I deserved it before seeking help. I used to tell her and others, and continually do, that help from above doesn't come on the first yell. The fruit most be ripe before being picked, or they will not be sweet.

A miracle only comes fast while flying because there is no time to wait, and everyone with wings has him as our copilot, but many pilots don't get helped because to receive it in distress, we must press the magic button not mentioned on the checklist: "*God*, take over, please!"

I felt a sharp pain in my right side by the belt, and as it progressed, I went to a kidney specialist. He gave me medication and the address of a hospital if the pain worsened. Passing by Caldwell Airport, where my Cessna was resting, I couldn't help my aviation weakness but decided to go for coffee at Lincoln Park, just ten minutes away. Temptation has a price to pay.

Two hours later, I took off, heading back to Caldwell, and the minute I was at only two thousand feet, I began gliding. I pressed the microphone button for the control tower to land, and then I began to yell, as I had a terrible pain on my right side. They responded that God would help me, and I had runways 22 and 18. The wind was calm, and their prayers were with me. My file showed I was a survivor, and the ambulance with help was there near the two airstrips on the taxiways.

I felt they were helpless to help me, but they had the powerful weapon called faith in a Creator. That is all we need at the moment of tension. Death is watching us as our time is coming. I noticed the fire trucks with their bright-red color were alongside as angels in good faith.

After a not-so-well-planned good landing, the waiting doctor gave me a shot of morphine as a gift from heaven, and three days later, I left the hospital as the same doctor said if it wasn't for God, I would be talking to him. He told me next time to listen to the advice, because it saves lives.

I affirmed to everyone I was going to leave my short aviation career not due to so many close calls as defying the Creator, as a show of machismo or foolishness as a pilot, note 10, but once again, I lift off. My wife was alongside to help me in my prayers, and coming down, she was going to add all my flights. I had promised Carol just before her death that I was stopping at my one thousandth landing.

Under so much pressure, I gulped and ignored her advice, but I was one of those who had lost any fear of flying, as an invisible champion, and according to the piloting philosophy, those are the ones who die first.

Here is another good one about miracles. I was around when it happened in New York City. A painter was doing a job outside of a window on the sixth floor, when he fell from it, going straight to the sidewalk, and gravity doesn't play games. His back landed on a rigid, thick cord, tied strongly to two posts, hiding ten feet deep in the concrete, and his body whirled around it. As he got up on his feet, he looked at the horrified group and said loud and clear, "It pays off to go to church every Sunday morning. I asked God for protection, because I am the breadwinner of a large family full of kids."

I shouted at him and asked if he had insurance. He said it was the best; otherwise, he would have landed on the rope, and then he went home for a well-deserved day off.

That was about five years ago, and it is very interesting, because I am writing this book as my last recounting on miracles. It is about my eye vision; at the moment, several ophthalmologists say I am defying science, as the spiritual world is not in the packet.

I noticed my vision was deteriorating, especially at distance. I was unable to read the road signs, and at night, it was impossible for me to drive, but wisely, if it was legal, I bought a seven-inch GPS, and with voice direction, I could pick and go anywhere right on the dime. I went to an optician and asked for an upgrade, but it was in vain. He recommended going to a professional for evaluation because he noticed I had a cataract.

The next day, I was evaluated as needing immediate surgery, but I kept rolling with time, waiting until the next month, and now it is passing the five-year mark. Around the retina, the milky ring is now a light blue, but at the examinations, I began reading all the letters and numbers and laughing at the doctor; she wanted to know the joke. I said loud and clear, "I am reading. I am reading."

She realized it as she saw how bad my cataract was that I had all those years. I was using eyeglasses only to look at small letters, but with lots of light, I could sometimes see it clearly. I am not challenging God's law, but I am a human being, looking at the spiritual world as if we were not spirits, but we better be or there would then be an eternal life. The cemetery is only a deposit of carnal bodies, as it later becomes a bone deposit. It would be a great show if Hawking decides (if he is around, naturally) to put me down for my remarks about him and it's proved I am reading with spiritual vision. I also hear well only with one ear, but I can hear a fly landing in the next room, not as a robot, but as a son of the Creator.

As I looked into the sunrise from my beachfront apartment in Copacabana, I admired the colorful clouds and was amazed as always in how much beauty is in our world of God's universe, when suddenly, a flashing brilliant light in both eyes blinded my vision. It seemed like I was in hell. I closed my eyes, and the blinding silver lights continued for a few long seconds.

I called an older friend, and he gave me the phone number of the best ophthalmologist in the high-class Ipanema Beach area. He said she was the best on the planet, because it was the movement of the vitreous or the gel that fills the inside. It was as white as an egg shell, and one of his friends was blind because of it.

I called the office and was told the next appointment would be in six months, but when I mentioned the lights, they told me to get into the first taxi, because maybe the doctor couldn't save my sight. I jumped into a cab and the taxi driver joined me in praying to the Lord, saying I was not ready for blindness. I would not be able to read and write for him or to make a living, because unlike Ray Charles, I am a bad piano player.

She took small X-rays and showed me the few millimeters on the top inside of the eyes had moved down. Little floaters, like small black specks or tiny spots, drifted around like mosquitoes. She said they were cells from the movement. She told me any ophthalmologist would say nothing could be done, but if I took lutein and ate greens like a rabbit, I could be the one in ten who was healed. I had to involve God in the process.

When I told her about my books, she looked on the site and told the secretary to buy them. Embracing me, she said God helps when we deserve it and I was sure a strong candidate.

Two weeks later, I was at a New York eye infirmary, and they told me there was no cure for my problem, and in less than a month, the annoying mosquitoes were resting in peace at the bottom of the gel. My eyesight was upgraded from two to a divine eight and all free from heaven.

She told me my cataract was like a cream, and it was a miracle I could see. She said it would be worse if I didn't change the lens immediately, because it was already bluish. There were two surgeons at the moment ready to start on the first eye.

Well, I decided to challenge man's knowledge but not God's wisdom. I have been holding off on the surgery for almost six years, while friends asked me if the light-bluish contact lenses matched my rosy skin. I have faith it will last to my nineties, when my galaxy XLT ship will take me back home, just like on the Star Trek TV series, which had no sex and mild violence and was family entertainment.

Because my trust in our Creator is very deep, as deep as it can be, I decided to wait for his decision. Please don't believe I am a religious fanatic, as it is a blind faith. As my American driver's license and also my Brazilian one had expired, I went to Rio's office to get one there. I needed it in Miami to rent a car at the airport. I knew I couldn't see the letters even with my glasses, but I believed blindly a miracle would happen, as it did. As the examining lady told me to read line 4, I told her to hold my book *God! The Realities of the Creator*, and to look at the last pages for my shoe and sandal designs. I whispered that I couldn't see well, and then I left feeling guilt while holding my license but more spiritual than ever.

A month later, in Fort Lauderdale, Florida, I took my eye examination, as I had my eyes on the numbers and on the examination lady. She said I was not using my glasses, as stated on my license, and then she became busy with the magical book, just like the one in Rio (Brazil), and I drove away with two driver's licenses but short two books from heaven.

While writing this book, I began on the *Grand Holiday of Ibero*. On the twenty-fourth of February, I got on the *Costa Favolusa* for seven nights to relax and stare into the sunrise and sunset, and during the wee hours, I looked at the pitch-black ocean and felt the difference of the dark matter, as it had a glow. I thought how good it was to be intelligent and be able to enjoy existing eternally.

Since I was born, my life has been a package of miracles or phenomena, but few would bother to know, because it is private to each one of us and that is what counts. All the glory is our spiritual passport and is not to put anyone on a pedestal. We are going to need it after we pass the door as spirits. As we are born from a microscopic zygote and then dissolved in the name of aging, the world keeps rolling. The microscope proved that even a rock grows. Everything is a miracle or phenomenon.

People ask me why I am so alive, sleeping few hours and never stopping, while reading, writing, talking, swimming, piloting, always doing something, even while going from New York to Miami and only stopping for gas and food but listening to classical music or great priests on the radio and laughing all the way down or flying for ten hours, but landing often for fun. My answer is: "A certain thing is to be born and die. Why should I sit down and wait for that moment that will come anyway? Consider it a waste of time because the monotony and immobility only rusts the body and soul. It gives time to age, because our minds don't stop. It is the spirit, and like God, it doesn't need to rest, because it isn't made of flesh and bone."

When you reach eighty years old, if you have slept eight hours a day, the total will be 23,360 hours being alive but inactive out of the 700,300 hours of existence, but if you do with only four hours a day, it will be 11,680 out of 23,360. It means I am learning and enjoying 11,680 hours more than you, and it is why many of us know a little more than others. It means being in class more, because our body is a school, or nothing else would make any sense.

Hawking affirmed that because of his physical immobility, he has all the time to stare at the monitor without distraction and discover the mysteries of the universe, but he must go behind the egg cell of the big bang research, because the dark matter is the nest. Pieces from the bang keep expanding into infinity; otherwise, they would hit the barriers of the mediocrity. One thing I don't like is to be known as a "brilliant mind," because it would follow any mortal to eternity. If it is without meekness, the light will not reflect out of the dark tunnel that we must face in the beginning of the long, unknowable journey beyond earth's crust.

When it comes to food to nourish our bodies, we are sophisticated because we are not irrational, eating what for us is tasteless. The chicken gulps down corn, and as its stomach becomes full, it stops. The dog falls on his back and sleep for hours because of a full belly.

As my answer is that human beings are born with the DNA for beauty and taste. I am not from Hawking's monkey family. We are the owners of the universe, because we don't just fill our mouths with food and gulp it down as monkeys do to bananas. We savor it, and we go now beyond the frontiers of, as I call it, the art of eating. All the time in my place or in a fine-dining establishment, I take my discreet digital camera and take pictures of the great dishes. At my monitor, mostly in the background, is an incredible dish from a showoff chef, as I am one.

I don't feel comfortable and even look with pity, when I see a human being eating grotesquely, as it reminds me of the irrationals. I had and have the pleasure of inviting people to my home for a great gourmet dinner, and I, as the chef, have had the unpleasant surprise of knowing they are not upgraded for a fancy dinner table. On cruises and restaurants, I often ask to be sitting miles away from my present table, as my right, and no one from the staff ever asked me what was the reason.

We were created to enjoy and appreciate all the creation for a marvelous eternal existence but with the free will to see it, as long we are here on earth.

The variety of products for our culinary taste is almost infinite, as are the methods of cooking it, but then I hear from people saying that life is boring, because there is nothing new in the kitchen. I find out they mostly eat boiled rice, pasta with oil, and so on, because laziness is still strong on the road of evolution.

Rice and beans dominate the Latin countries, while it is rice in Asia, potatoes in Europe, and pasta in Italy. They have been adopted by everyone because all they need is a pot of boiling water. In the Arabian world, they love to lick their fingers just like a KFC advertisement. I do too when eating boiled blue crab on garlic and herbs.

One important thing is to eat two boiled eggs, minced with olive oil, ginger-roasted seeds, and fresh chopped parsley daily, and it will give you all the energy and protein to follow me around for twenty-four hours. The latest report in medicine that egg yolk raises cholesterol is untrue, as margarine is not healthier than butter. I am eating two daily and drinking milk from a bottle. I eat fat from meats, as margarine was always out of my diet. As I said, I wasn't born to cook for the hospital and I eat everything. I cooked to perfection as is our right as human beings with a certain grade of culture from evolution. I have been holding well for eighty years, keeping the doctor away and missing my appointments, making room for the ones in critical need.

If you want to lose weight, do as I do: eat everything your heart desires, but cut the portions in half and make the salad as a show of beauty, including slices of mango, pineapple, orange, or any fruit on lettuce leaves, with a touch of any dressing and always include Portuguese olive oil, a dash of apple vinegar, and sprinkle it with fresh grated parmesan.

When stress takes over, it's free; it could be daily, as it comes from any direction, as at the job or business. Don't run to the physiologist or God's man, because he or she could be in a worse situation than you. They are human beings and not angels. Just take a deep breath and after 10:00 p.m. go for a long walk away from people, staring often into the sky and being grateful you can think. Seek only the good moments from the past, because they are never lost from our memory bank.

Once in a while, pick up your coat as you are going to put out the garbage and go to the cozy waterhole nearby as it is always full of happy souls. Ask for Oysters Rockefeller or a hot pastrami sandwich on Jewish rye bread, heavy on German granulated mustard. Gulp it down with fresh coffee. Sip on a glass of dry red wine or a glass of ice-cold beer—having only one drink—while watching our brothers and sisters, laughing and talking nonsense, and you will not feel alone in the universe.

Minutes later, you see that all the anxiety is gone, but if someone who bothers you comes in, leave and go to the nearest coffee shop. Ask for a scoop of your favorite ice cream, and at the counter, talk softly with anyone. Coming back home, you will be so light that everyone will ask you if you have met God, and say, "Why not?"

Anywhere on the planet, men go to a bar and ask for coffee or a Coca-Cola, while others get loaded on heavy liquor, becoming candidates for accidents, losing jobs, and divorcing because alcohol goes straight to the brain where the spirit waits for disaster.

Years ago, when flying first class, if you asked for a glass of wine or whiskey, the staff would put a bottle alongside you for the duration of the trip, but now, because of moral evolution, the crew will ask if it will be a glass of red, white, or rose wine. If you call for a refill, they will bring a nice tray of appetizers, saying the booze supply and the wine case had broken during turbulence.

If the pilot or copilot drinks a glass of beer within less than forty-eight hours of flying or, better yet, two days, he will be asked at the airport for a blood test. Naturally, someone whispered it to the authorities, and if he refuses the alcohol test, he won't be permitted to fly the aircraft. Days later, the FAA board of judges will clip his wings for good, because God will never take over in any emergency if the aviator smells of booze.

In my houses, apartments, cars, planes, business, and even backyard patios, the use of alcohol was limited, smoking signs come in every shape and color, and the violator would get one warning. The second would be to leave on the spot, and if they resisted, then the police would help, because we all have our rights to live decently and not carry someone else's cross.

I noticed many years ago, when we sincerely believed in miracles and phenomena, that the ones who did received them more frequently than the ones who blasphemed, because it's the effect of the contact of faith in the Creator. We also must deserve it, because nothing comes from nothing, including healing and happiness. To be born or not to be born; that's the mystery.

No one will ever know, because it is a question, not available to anyone here or eternally, as part of the Creation. He holds the key of wisdom.

12 To Be Born or Not to Be Born Is the Mystery of Existence

Practically, even the incredulous says it is the number-one phenomenon. The religions glorify the Creator as a miracle, but I don't agree, because everything is a miracle or phenomenon, including the air we inhale and the *276,480,000* heartbeats I have had in my eighty years, with my heart still pounding perfectly, pushing blood through kilometers of tubing of veins and arteries without stopping.

The astonishing marvelousness is it is made of flesh, soft as gel and strong muscles and bones, as strong as steel, with an array of tubing and its beat is perfectly perfect, using an electrical sign, saying to those "brilliant minds" to stop in their tracks and look inside themselves as a perfect machine, fueled by natural nourishment that grows from the ground and irrational carnal bodies that we eat.

The ones who say everything came from ground zero or nothing, originating itself from a nullity, did not came from zero ground, but the Supreme Intelligence, and they make a mockery of themselves. As observed, the ones not classified as "brilliant minds," are not at the top of man's society with trophies, but that doesn't mean they know the truth above others, because sooner or later, time will roll into the future and will show not even dust from their bones, but surviving eternity are the good deeds and wise sayings. Not everyone is on the same level, as high systematic universities can't teach wisdom. Insight in each one of us is already planned at the birth DNA, according to the spirit.

Our creation begins deep in the radiant dark matter, as did the egg of the big bang, and continues right here between a man and a woman or a male or female irrational on the ground or in the water. Life begins when a male is aroused and at his climax ejaculates unconditionally or naturally and the sperm meets an egg. The resultant creature is measured as being as small as a micrometer (one millionth of a meter), having a mass between 0.00177 to 0.0042 milligrams. The ovum is the perfect diameter. The ovum cell is 0.15 to 0.2 milligrams and the radius is 0.075 to 0.1 milligrams alive. We called the male gamete sperm, the name from the ancient Greek. The mature male cell as a germ has all it needs to enter the perfect round female ovum to create a zygote, and then his or her growth begins. (I use the masculine and feminine pronouns, because we are already alive.) Normally, a few months later, the mother can feel his or her movement. Nine months later, at the moment after cutting the umbilical cord, he or she yells, and bingo! We are here to begin an amazing existence and do what we please as free will.

Also, it comes into the package. As we make our path to infinity, we can do it gloriously or not, because included in this marvelous system of creation is a mind behind and above it. Of the ones who do not accept it, I demand an explanation. While not all agree how he does his work as a Creator, those who go to a point of denying his existence have the "nail" ready for them. Many tears of happiness and sorrow have cost me thousands of hours, and eighty years is considered a long life. On those sleepless nights, not at a convent, seminary, religious university, or pompous college, I have had my bed made, gourmet food, and sacred music to soothe my soul. Among people everywhere, they feel it in their hearts while facing Satan and his affiliates, but they do so with the help of angels, who come here not disguised as Stephen Hawking, but as William Moreira (Canno) and many others.

We have Mozart, a genius at birth, but just in one degree, the degree of classical music. Walt Disney was in entertainment, and his creations contributed to our happiness at all ages; he was a genius in cartoons. He had a mind of beauty, as did others in several degrees, like Thomas Edison who created hundreds of inventions the world needed so badly. Each one is on a mission and destined to help improve the earth, but there are also those who are able to be in all the degrees, but it's not enough to know and feel the grandeur of the total and not just to scratch the surface.

I mention the positive side of the coin, because on the other is a contrast. It is the negative one. The ones who come programmed to awake us from the good and the bad as necessary in our existence or the universe itself.

No human being today has ever dedicated his existence exclusively to denying a spirit as a Supreme Intelligence. While rewriting this book, I watched several times on Discovery the show broadcast in August 2012: *Stephen Hawking—The Grand Designer: Did God Create the Universe?*

Those who have dedicated themselves to helping humanity have never been called "brilliant minds" or accepted this illogical criterion, because this expression elevates a human being almost to the Supreme Intelligence, but for those not yet at a level of intellect, it makes fun of the ones dedicated to helping society. They are on an illusionary red carpet, which impairs the logical spiritual world, and retaliation comes in as catastrophes, including physical and moral, to those who are kept erroneously on a pedestal of witlessness, as a Roman plot to destroy the ethical system of a decaying society. Soon, what will be left of it is the whispering of time, as it creates the past, where all the lights are now off or going off.

The only "brilliant mind" is the one ahead of time, as *the Grand Designer*, where there is no logic or common sense for the public. There are notes of papers, giving no direction, not even to the beginning here, in our material world. We come from a zygote as our big bang, and then we grow on as our own universe, because we now are part of it. We are created like an atom or a grain of sand that will never disappear. As Lavoisier said, after things are created, they will never be lost, just transformed, as he now goes from soul to spirit.

This goes for everyone with or without the Creator in their minds, because they are created, and it means there is no power of anything but to exist eternally, enjoying it as it is, because he put all that is necessary at our feet.

Some scientists—could be you or I—somehow eternally know nothing about the beginning of the existence. We will never know about the dark matter as the microscopic or giant egg explosion, like the big bang or the birth of the universe or universes, much less how he had his origin, but we know our carnal body begins right here at home, just like any animal's, as we do artificial insemination of the cow we eat as food or our children whom we love eternally. The process is the same, and I believe it should give us some food for thought, as being graceful is not being a classical breed of monkeys, as affirmed by Stephen Hawking and backed up, seriously, by the media. They declare him a brilliant mind, but not as a comedian, like Bob Hope (2003). He did not scare us but provided enjoyment at its best, living one century healthy and happy as he loved the world and we loved him. He always had God at the tip of his tongue graciously.

Millions of disabled people are frozen in time and space, without freedom to walk, run, and fly, while learning continually that life is beautiful no matter what, are enjoying it as they can and waiting for the eternal call of a great free life so incredibly real, as the way we came into being. They are our coming to this miraculous carnal body, as our step to eternity.

Logically, we are more than this, more than a letter in the alphabet, as one begins, but not an end or there wouldn't be intelligence or hope. Even without seeing the face of the Creator, it is better this way, because no one could judge him as beautiful or ugly, tall or short, black or white, as it is done about the image of Jesus, which some believe is discriminatory because some black people believe Christ should have been black. Should there be several Jesuses—Chinese, African, Caucasian, etc.? What color of eyes? As it is, we can play with our imagination while he keeps his spiritual image, as the sky has no limits.

One simple example has happened to all of us, especially during our early dating process seeking a companion. We could talk on the phone imagining him or her as the right one, but when the time came to meet, we would run from it or maybe both, and God as the Supreme Creator would like to keep us wondering eternally for our own benefit.

13 To the "Brilliant Minds," Where It All Began

This quotation, as any brainless person knows is God's dilemma, because he is the one who has all the ingredients (atoms or whatever Hawking has dreamed of) at his hand to make another zygote with a soul and karma, destiny, and a mission already programmed.

He wishes to "break his chops," as he then shows all us that no one is above him. In the early mythology, the Vikings, the Greeks, and even the almighty Roman Empire had as many gods as they had soldiers. The Catholic Church then had their successors come up with hundreds of saints, but saints are just God's servants, demonstrating evolution, it walks like a crab sideways but then forward, as the Romans became Catholics, putting together the god of the sun as the ring with the cross, and spiritual evolution leaped, almost reaching heaven. Christ was announced to the four corners of the planet as the symbol of love. His parables and the Sermon on the Mount are all atheists need to get up from the wheelchair of incertitude.

Someone officially deserves the Nobel Prize for Peace, and that is Jesus Christ, but not postresurrection, because he is the only one alive in everyone's minds, not as just a Bible character but now ascertained as the number one in everyone's hearts, even the godless people, when the firing squad aims and the call is, "Fire."

Before the bullets explode in the heart of the sinner, Jesus's name echoes inside the prison walls, and during the French Revolution (1799), the drums were beating loudly as the guillotine blade was obeying gravity to mute the last condemned yell to Jesus.

This time, you are wrong, because this quote: "*To be born or not to be born, this is the mystery*," is not by William Shakespeare (1564), but William Moreira (Canno) (1933), because in his epoch, science was exclusively the province of the Vatican, having angels and demons under their feet, as Galileo was not recognized. Yet, to the Vatican, the Nobel Prize in Science was not for the great astronomer, mathematic genius, and so on. Sir Isaac Newton (1704) had his concept of the reflection mirror for a better telescopic, and I believe this is because he didn't show up for Sunday's Mass.

When I mention religions, it is with respect, because they attract almost all the lost lambs, and it has a great array of choices to pick from. Because of that, faith and hope are alive and well, keeping order in the flock with huge numbers under one God, the Creator and nothing above and beyond him—in general, humans agree that there is an intelligent power behind the existence.

Imagine! If the plot of Stephen Hawking's godless existence had worked as he kept pressing as true his theories, but ignoring the beauty around him, being frozen in a body. The media loves Satan to glory in the inequality, as bad news always makes headlines when money is involved.

I wonder if Hawking would go back in time, as warping the present landing at the inquisition epoch and facing the godless theory of a existence coming from a zero concept of nothing, his future would have being roasted to nothing, sparing me of having read the senselessness of his nonsense dreams, fantasies, and thoughtless quotes and having wasted thirty-five dollars. Then I lost two hours reading *The Grand Designer* (2010) only to have it be the first book of thousands I had read in my short eighty years I trashed. I danced the samba and jazz to distract my soul from stress.

Once more, I felt the philosophical thoughts of a man condemned alive to be a mummy, while his brain and communication are perfect, thanks to someone with intelligence doing his great mission to humanity and not making theories. The great Bill Gates saved him from a heavier cross, as a real monkey with no talking, but listening while eating bananas. Bill Gates has a normal life giving us facts, not black holes. He did not kill the hope that belongs to the poor and rich alike.

He is expressing his contradictions, going against all odds, passing the half-century mark, swimming against the flow of his birth on the tide that glorifies the Creator. It is his choice in free will, and like it or not, he is paying with a heavy cross, but with option, which millions had hoped would appear in *The Grand Designer.*

But it was negativity as he rebutted his last chance to be free of the ordeal, I do not wish it even on criminals. Besides, the ones siding with him in all his thoughtless fantasies and dreams bring nothing practical to our challenging world of ordeals, while surrounded by colorful beauty and love as human beings. We all try to demonstrate by music involving our soul and spirits, as religions try to give us words and evangelical lyrics to console us, especially when tragedy strikes. No one is spared, as everything has a reason on our step to eternity. I lost my beloved daughter Carol, as Hawking went to his wheelchair to keep his mouth behaving, but perhaps he loves to suffer to cleanse his soul. I will mail him my book: *How to Suffer Happily* (2001).

I began following Hawking's path as early as his twenty-first birthday. I was thirty-one. I have survived a half century glorifying life and accepting all the rocks that hit me hard. I glorify in the moments of ointment given to me as a consolation, especially seventeen years ago when I faced seeing the body of my daughter in pieces at a New York City morgue, and religious leaders from all faiths, including Muslims, ran to embrace my hot, bitter tears of disconsolation, giving hope. Hope is the only consolation that would soften a heart fragmented by moral pain, especially of a father or mother.

If Hawking was around to tell me not to worry because now Carol is inexistent as soul or spirit and would be a pile of insignificant atoms, I would have guaranteed him to have a life mummified, for sure.

He, with his tongue, had lasted his half century tied to his own ignorance. He blames nature for it, absent a Creator. If nature was involved without spirituality, a rational society wouldn't be possible, but hope must stay around for all of us. Even passing into my eighties, I need the concept of a God to bond my life as we all do; otherwise, there wouldn't be any respect and suicide would occur more often. Being born is not a choice, but terminating our life here is an option of the free will. But there is a high price to pay. It is like a person condemned to a prison sentence jumping the wall; when brought back, the penalty will be double. It is our option as pain and problems take over. Sooner or later, it will make the skeptic say: "Oh, God," and then get up from the "divine nail."

He says it is not necessary to have a Creator for our existence. He tries hard to make an "obtuse" person of himself as being from a faraway galaxy. He is telling me he knows more than anyone here, as he stepped in the minefield, not belonging to my club. He ignores the laws of common sense, as part of the intelligence, but it is not officially his fault. The dull-witted ones look at him as having glory for themselves, but he is the one paying the high price.

If he were a "brilliant mind," he should have used the light, as I use mine, to seek the one who built a perfect zygote called William Moreira (Canno) with all the quantum theories necessary for me to face my eighty years—running, piloting, writing more than one hundred words a minute, enjoying the world, and with spiritual contacts affirming that life is more than being born, suffering, and dying. It is to enjoy and not be sitting on a "nail" as a punishment, wasting my chances to get up, not as a "brilliant mind" but a "thickhead."

What good is it to be praised by society if I can't enjoy anything, as payment for blaspheming a Creator? Even an uneducated street cleaner in a third-world country knows better up to Einstein, who had glory, and they all lived, walking and doing what you and I dream of doing.

Hawking was invited as the entire scientist group not being part of the Vatican's private science reunion in 1985. Later, talking to a Catholic cardinal in New York, he said they tried to tell him about the grandeur of our God's existence, but they lost the battle to save his body and soul.

If someone tries to make me squeal for names, before and especially now, my answer is with the aging process of the body, everything is kept in the memory bank, except names. If someone eats lots of cheese, like I do, the scrubbing process accelerates.

He should abandon the unreached infinity, about which he knows less than any doorman in Brazil, and dedicate himself to the mystery of the zygote phenomenon where we come from and find out why there are so many millions of sperm being created in just one bang, all alive with energy to fight to reach the ovum as the race for survival already possessed by a spirit (us); otherwise, there wouldn't be a race, and then the others that didn't make it, just die or go back to the line.

Those little specks of germ cells I dare to call our carnal and maybe spiritual beginning. (I dislike calling them theories, as I don't belong to a world of fantasy, but facts.) It is clearly divided into sides on the Internet, but as a "brilliant mind," he would be able to do autopsies on their brains.

That means the sperm, with a head almost half of its size (I believe his sperm was the entire size of a brain). It must be called he or she, because it is us. We are rational.

Hawking disagrees God created us as primitive animals, but humans in evolution were already monkeys. When tired of bananas, we began cooking gourmet food, trading a tree-to-tree travel for jets. He reminds us of the days of his family monkey business. That's what happens when you play with fire. Sooner or later, we get burned for sure, by a William Moreira (Canno).

The sperm is reality. It is under our nose, more than his black holes that are untouchables, except in his "brilliant mind," because as it moves to reach the ovum, it needs energy and concentration, and the Creator wouldn't mind. We do a billion autopsies of sperms, because only one to eight reach the ovum, and the others are already condemned for the trash. The Vatican would interfere; that I can guarantee.

I am sure he can do that and be more applauded by his fan club, as I notice our President Obama is part of it or was forced to be when he invited, not long ago, the number-one godless human being for a medal decoration but wasn't invited for dinner, because there wouldn't be any manners at the table. Everyone would be staring at their plates, eating as fast as they could without waiting for coffee and dessert.

Human life forms here have intelligence, and only hobos don't come from a sperm and egg, but from a machine shop, not up there among the galaxies. To be born, we need a male and female, and not an immensity of space with atoms, because it gyrates heaven but not intelligence.

Science must begin from the roots and not the treetop to water, feeding the vegetable world, or it will wilt, dry, and decay as has been happening so far, wasting time researching answers too far away to reach.

Aristotle, in his epoch, did not have one god but many. He did observe the shadows of the moon and the earth's eclipses. He used a stick to make circles in sand and to figure it out as he took a small step. Because of that, Hawking said he is his number-one admirer, and meanwhile, he ignores his free will to become disreputable against all the odds. He fights back against the one who created him. Aristotle, you, or I would not be a pile of sand but of atoms (the beginning of the division of the invisible world of the matter).

In 1979 and 2006, I went to the movie theater to watch *Black Hole, the Movie*, interested in how far Hawking's fright had affected his dreams and fantasies against the creation. People took it as Hollywood entertainment, another idea of Disney. Otherwise, they would have been running around like headless chickens in a circle. I seek him to give a medal of morality.

As stated by the "genius," our brains have about one hundred billion neurons. I will give all my paints and designs, my collection of classical music, and my beloved *Parker 51* that helped me earn my journalist ID at the age of twenty and throw in as a bonus my Cessna and my blessing in the name of the Creator if he could begin such movements. Also, if he tells me how he came to his amazing total, and even more, if he tells me in my ear if he is afraid of death, and that I will keep secret.

The "brilliant mind" may come to this brain theory while doing an autopsy or an anatomy of a sperm, while in a class of quantum theories, seeking the soul confirmation of an afterlife, but cadavers are soulless or medicine would have to give a general anesthesia to cut down the pain. We feel our pain while in the body's brain, as a way for our creator to punish or alert us of something wrong with our spirit. I hope some "nuts" (not Brazilian) surgeons, while working inside the brain seeking a tumor, did look closely to see the soul!

As amazing as everything in the universe is, we all swim into the uterus, never out, even Hawking, as the Nobel Prize winner for being the only one who dedicated his whole life to not having the opportunity to seek spiritual help or to suffer less, but his career as a godless man to be at the podium of an eternal being. He did recoil when he picked as the base the number zero and the word *nothing* as the rock foundation of his legacy, because it was sand. He has a life of people looking at him in sorrow and not as a genius. As to those in his position, his ordeal is beyond the understanding of those not deep in spirituality, as I am.

For the making of the universe, he said God was not necessary. He feels being the Creator requires only two ingredients. One is genius, and now after the first one went to heaven (Einstein), he stated it was necessary to have three elements, but later, he said it was a mistake. The second genius took his place, affirming as he does with everything out of his mouth, that there are only two elements. Meanwhile, the Creator creates the flour, yeast, and also the bread. He began like anyone else from a microscopic zygote, or he would be around. After a while, the bread as his glory will end up in the oven, as his choice: cremation or being eaten by germs. The Creator, as the pilot of a flying aircraft, is the only high authority on board, because when something goes wrong, we are the only one to take actions and the blame as the ship's captain.

As space and energy came from *nothing*, then *Hawking* must give the formula to save the planet. It should also include grains and clean water, but he better hurry up, because the Creator has his eyes on him. The departure time is around the corner.

Before leaving our world to become just a pile of atoms, Hawking must find the solution to his quantum idea. Finding the smallest composition will have me lost eternally, because there is always space between any composition, even if they are back-to-back with each other. There is space in the between, and it is filled with questions that only God can answer for me, but he does not have to. He wouldn't; otherwise, he wouldn't be God.

Religions give to every soul without asking for our university diplomas, without filling us with senseless equations and theories, because 95 percent of people would understand. The majority are not based in theories, as like a fantasy of the ones who know nothing that is not in front of their noses but a hope that there is an afterlife, based in love, bringing consolation to the broken hearts, as I felt when my Carol left us.

Now, science has the menace of us being hit any moment by an asteroid that killed the dinosaurs a few millions years in the past. I wasn't there to confirm it, but black holes as big as the sun or bigger by a billion times swirl around devouring even light. There is no way out. Today, the genius, as a "brilliant mind," is sitting on his "nail." For over a half century, he has been defying death in a body. I beg the Creator that I never deserve this, as Jesus begged to avoid the suffering on the cross. He also said he is the one making the choice and accepting it. Hawking did not accept his cross and believed he could run from it as denying the Creator, but no one can run from his destiny. It is created at the time the sperm is just an iota, swimming to enter the ovum.

On the National Geographic or Discovery Channel, we see people born without arms and legs, just with fifteen-inch stubs, having a normal life, laughing and enjoying life. They drive cars, work as businesspeople, and do more than those with full arms and legs.

Jesus is a legacy, and it's more than an icon, because his name is pronounced daily by millions. When the atheist macho men faced the guillotine by the hundreds a day during the French Revolution, when the drums stopped and the blade began descending, there was no way out. They had one second to repent before darkness took over. They were screaming or whispering, "Jesus."

As for the brilliant minds or upgraded geniuses, their names will fade away as fast as the grave is filled with dirt, and the next day, headlines will come up, and as the wind of time whispers, the next generation will not pronounce their names anymore, because the legend was nothing but zero, because theories begin in kindergarten. Now let's make believe there was a monster and an angel.

The effect of this illogical or satanic propaganda provokes depression without healings or solutions, in a world. Sometimes we feel we are sitting in a volcano, ready to erupt, like the people in Naples (Italy) where Vesuvius is nested, ready at any second to do what it did to Pompeii and Herculaneum, as well as Stabiae and Oplontis.

The present inhabitants live there knowing Vesuvius is just dozing off, ready to begin again, as it did many times but mildly. The big one could come at any time, as the end of life. Everyone there thinks it will only happen next century. As I was vacationing there years ago, I asked people why they do not go elsewhere. The answers came almost logically: "No place on earth is guaranteed to be safe, because the plates are always moving, provoking earthquakes, tsunamis, and so on, and worse are the atomic bombs ready to incinerate us anytime. North Korea and Iran are working hard to belong to Satan's Club as super powers. We are all powerless. It is the effect of the cause. Science is proud of it, while the medical field is set aside, though we are being eaten alive by microbes."

After that, I kept my mouth shut as everything had been said, because not even the right ofbeing born is guaranteed. Some women come not as angels, but the devils of death, as they abort a life of a child from the uterus in the beginning, as it was just an inconvenient germ, not taking an antibiotic, but a pill called "baby eliminator," and if that doesn't work than a spoon-knife will do the job. That's why one in every 150 to 200 births has a deficiency.

Even children in kindergarten know what is in a fowl egg. Many know they also came from their mother's egg, as the brother or sister was born. It's all right, because children listen in the right ear and let it go out the left one.

Here, we have the number-one mystery, right from our feet up to our noses, and we don't have to travel or spend on research, because every man and woman carries this fabulous chemistry of creation as the beginning of life. Too bad I didn't write this book fifty years ago, so Hawking wouldn't have stared at his monitor. He would not have sought black holes, but all the secrets of the sperm and ovum. He could have found God, or at least gotten closer. That would have been more than enough for him to get up from the inferno of the wheelchair and enjoy the world as I did.

When science begins discerning the brain of the micrometer germ cell, as he or she has a head many times bigger them the whole body (just look at the sites), they will find a clue for the "brilliant minds" to find out how many neurons are in it and compare it to us as adults. There will be no shortage of supply, or any complaints from the human rights organization, as no zygote ever put up a complaint. The legal age to begin a legal protest is eighteen, and by then, the complainer would have to account for his own production line.

Everything in our world should not be looked at as insignificant, as we are above others as unique, or even on the top, looking down at others. We are very important, as acclaimed by the multitudes, because as a blind person, we would fall from the pedestal to a nonlife.

Losing one's eyesight would take anyone to a world of darkness, where he or she will be lost in need of someone alongside to not let us fall into the next hole. A dog can help, but that is not the solution, because we will not be able to take care of the animal. If someone has just become blind and has access to a gun, he or she will just put a bullet into the brain, as I found out with some people in my life.

To be born blind isn't great, but what we don't know we can't miss. For them, there is no choice, as Hawking should have been grateful in not being blind. Otherwise, the only solution is playing the piano, as a friend of mine in Rio de Janeiro did. He died four years ago at the age of seventy-six. He was blind for seventy years, but he became a great pianist. He told me smiling that he felt God all the time when he sat down and hit the keyboard. He got lost in the marvelous world of sounds, knowing the applause of the public that always surrounded him.

He told me many times he was particularly proud he could go alone to the bathroom and take a shower, because everything was in its place, as was the knife and fork. If anything was not as planned, he would know, as he could find anything. He was joking when the electrical power was off, it made no difference as long the refrigerator was cold. That was when he knew the power was back.

He was grateful to a merciful God, as he became an avid reader in braille, and, most interesting, he lived in a beautiful building only half a city block alongside the entrance of a great shopping center (Rio Sul) and also near my apartment in Copacabana Beach. He had a big, fancy piano at home and another at the center lobby of the busy mall. He could go and play anytime he wanted. He even walked alone to it. (He didn't know he had been tagged all the time.) He was pampered by thousands as he hit the first note. The applause could be heard on the sixth floor, echoing through the marble floors and walls. The grace of his life was he never complained of our Creator.

The piano is still there, with his picture on it saying: "Here comes music from heaven, from an angel, called Senhor Americo, as he now is back to where he came from, while we all have him eternally in our hearts."

He was known as "Senhor Americo, our angel."

A young lady, Isamar Coufal, was his companion at home, walking him to concerts and the innumerous social parties as she administered his apartment. He used to tell everyone she was his light, and after his death, she came to me offering her services in taking care of my life as a senior, because at my age I was family-less, as even my daughter Carol left me at the age of thirty-three, and it was seventeen years ago. I accepted her offer, and weekly we walked to the shopping mall to listen to the piano while in the huge hall. They have a dozen great restaurants—including Outback, Houston, California Pizza, fast food like Burger King, McDonald's, Kentucky Fried Chicken, Dunkin' Donuts, and innumerous coffeehouses, such as Starbucks—representing the United States and great Brazilian classical gastronomy well. The image of Americo will never fade away, as everyone looks at his photo on the top of the piano where the flower pot is always full of red and white roses.

As he was well informed about our world, he told me he felt pity for Hawking for not being physically blind, because he could teach him how to play the piano and reach the Creator, while walking around listening to the birds and the sound of the big waves at the beach and at dance parties. It is like being in heaven, because he could smell the perfume of each lady. When his health deteriorated, he told me he was going back to the world of light, because spirits aren't blind. He told me to go ahead with my *Parker 51* as my piano and the computer as my keyboard.

I had to give his example. We are surrounded by good and bad spirits, coming as music players, coffee shop managers, and scientists, scaring everyone's guts just for the hell of it, but each one is receiving what they deserve as the effect of the cause. Carol visited more than ten countries, and she visited the poor areas. She did well in languages and always gave parties at home. Like me, she was great at the stove. She especially mastered crème cheesecakes. She filled the house with at least fifty guests, and no one ever called her by her name, but always by a nickname: "My angel." She answered, "Yes, my love," even to Mr. Marriot while she worked at the Marriot Marquis in Times Square when he came to her floor for a coffee break, but everyone knew it was to see his girl.

People always say the good ones are the first to die. I agreed, and the reason is because they have accomplished their mission. It was then time to go to the eternal world and not stay here anymore, where the negative ones are glorified as "brilliant minds" or "geniuses."

One thing is for sure, when they all go into the spiritual world, they don't let those already with virtues fill out the card, because it is ready.

The title, glory, and money power from earth is nothing there, because the verdict is based in good deeds and not in royalties, Medals of Honor, or Nobel Prizes, except Mother Theresa of Calcutta's for charity beyond borders. The same goes for Gandhi, Martin Luther King Jr., and many, many others who only sowed seeds of love—they didn't have tostay in line.

Now that we have the principle of our origin as a human in a carnal body established, being materially perfect and spiritual in the universe, intelligence is destined to eternity. Our understanding of the Creator is he did everything the way it is; it is not our concern—you, I, and Hawking, as God's "chop breaker," and all the aborigines who only think it is the way to get their next meal. The alcoholic wants his drink, while the irrational keeps his nose in the grass, and on and on. It cannot do anything or, better yet, has to live as it is, knowing not of the laws of cause and effect. Do your part to enjoy it eternally or suffer until you learn it is for everyone not irrationals, as their existence ends up being our dinner or another irrational for a meal, but we intelligently float away from our diseased body, already with some preparation to face the afterlife from the material world of the flesh.

"To be or not to be, is the question." Now, I offer, "To be born or not to be born is the mystery." My answer to someone in the wee hours on the top deck of the cruise ship was this book being born. It whispered to me, while the wind was hitting my face at sixty kilometers, as to the atheists already condemned to be born or not be born, there is no future for them after death. Because of that, I as a pilot never met a godless man flying any airplane. If one is found among us, we clip off his or her wings on the spot and trade the plane for a car, because in an emergency, the Creator will not be at the right seat.

The first quote is our choice, but the second one is my gift to all of you, as I understood it. There is no choice, because the decisions are already made, not only for our existence. The path to follow is on earth, but with free will as our freedom, as with everything else, there are rules and regulations, making our eternal existence a busy one, full of ups and downs, but always colorful.

I used to suffer as a younger man, but as age takes over in our material life, the horizon of an afterlife becomes clearer as family, friends, neighbors, and strangers just pop out like a balloon in our faces. You know anyone could be next at any second, and as the miles are left behind, we know the landing is ahead, and the landing is a must. The crowd is waiting with the good news to direct us down a new path.

The ones who do not agree with me, use your free will to enjoy the godless scientists association as part of Hawking's Club, offering no afterlife to worry about. You will land in a cannibal party pot, in the guts of hell, as a fantasy to the ones who came to earth "to break anyone's chops."

I do not, and I never did stress or get deep in paranoia about the future. I accept it as it is, but I always try hard to better it, by helping others whether they ask or not. First of all, it is not to believe in death, as it is obvious to anyone. Seeing that we are born from a speck that is a micrometer, about the size of a mustard seed, to a full enormous tree. I mean the human body affirms our amazing origin, putting no doubts about it; there is no end after the beginning, because we are intelligent. If your body and mine fall into a pot, even if it was an accident for sure, cremated or entombed under seven feet, it makes no difference, because the spirit is not following it anymore. What counts is the spiritual fluid human form.

The movie *Ghost* was a successful Hollywood fantasy. It came to Brazil ten years ago. Millions of sick and disabled people had contact with the famous *Dr. Fritz* in Rio de Janeiro. One of them was Christopher Reeve (the Superman) as he became paralyzed just like Hawking. He fell from a horse at a show, but Dana, his wife (both deceased), demanded the medium Rubens to come to East Orange, New Jersey, and *Dr. Fritz* said the ones to be healed had to go to the saint, and both left with the proud heart as he died as a mummy in bed, and she was healthy but died shortly.

People were healed like in an assembly line (the *New York Times* had a special issue about the miracle of the millennium, as I picked the title and wrote 440 pages of a true story). Dr. Fritz was a spirit, working with Rubens's body as a medium was a spiritual showoff. Millions of surgeries were performed openly in minutes without anesthesia and disinfectant.

Francisco Xavier (Chico Xavier) wrote 420 books about spiritual philosophy and giving answers about the spiritual world. He came from a small humble city and only attended grammar school, working all his life at the post office. He never went to the movies or left his hometown. I qualified him as a born "brilliant mind" and not the fake ones from universities. He received the Nobel Prize as a peacemaker to the world. While his eyes were closed and his hand held the head, he wrote two hundred words per minute, and the line goes beyond the horizon. The spiritual world is allowed to communicate, but the average people did not pay attention, because their eyes and ears were on the ball game, idiotic shows, pornography, novels heavy with sex or gossip, as the planet is rolling toward destruction, only to be digested by Hawking's black holes.

Not long ago, the programs for everyone, such as Bob Hope, Lawrence Welk, and dozens of family shows, dominated the tube (TV), as soft music from the great big bands elevated even the irrationals at the zoo. Today, music feels like it is coming from the deepness of the sinister world, as it doesn't smooth our stress but accelerates it to a madness, like the ones jumping up and down, remembering the aborigines doing the ritual of death.

Stephen Hawking complained he never was appointed for the Nobel Prize, and maybe Satan could pull some strings for him, because religious people say the ones who reputed God did so because they have a pact with the underground dark world. The ones who never tried to have contact with vibrations from the energy are the unhappy ones, not as disbelievers, but as those afraid to have to bend their heads in humbleness because of their position in society. It can be accomplished inside the bedroom on your knees with a hand up above your head, and a heart with sincerity.

The ones who are the majority of the population are religious and have a better life, based on feeling the spiritual vibrations, not visible to the naked eye or in a microscope but they can be felt. In the cruise ship, I feel it in the moment that I step aboard in the buoyancy, even on the side with smaller propellers. It's the unseeing vibrations, but it's there or it wouldn't be felt. While aboard, the big propellers started turning. I know we are moving toward it, and it is like a blessing. Imagine if we could feel earth rotating?

The ones who feel nothing do as they choose. Spirituality is not part of their existence. There is a reason to get the seniors to fear a life of nothingness. They are afraid of the wrinkles, white hair or losing hair, and spots on the skin as aging is here to stay, and physically, they slow down, always seeking a seat to put down the weight of the years.

I was born or conceived in Belo Horizonte, the third city in Brazil, in 1935, but my parents affirmed it was in 1933, as there was a revolution brewing. They went to live on a farm a few hours driving on a dirt road from the capital, and two years later, back in civilization, they registered me as being born. It makes no difference to me, because I have the two dates to throw a party. One is on time, and the other is at a second time, as meaningless as the atheist's life.

The drama of science where time, space, and relativity are like a spiderweb, and only the spider understands and dominates it, because the spider is the one who imagined it and built it using the material of life itself, while everything and everyone else is just involved in it without a way out. The more there is movement, the more impossible is the comprehension of the complexities without being a divine spider, because it is the only one who can navigate freely without getting involved stickily in it, with a way out.

If you do not understand what I am painfully trying to put on paper, the logic of our world in God's Universe, then my conscience is free because I have tried it, and the rest will be history, your history.

You and I are the universe, including the hollow, infinite faint glowing dark matter, because without the human being to be in it intelligently, appreciating its glory, then all the existing matter would be nothing, like the world of Stephen William Hawking.

One day, it wasn't a day but time, as part of how the infinite faint glowing dark matter, the Supreme Intelligence decided to share his existence with a universe pulsing with life, but as intelligent beings to be in it, like him to appreciate his creation; otherwise, you and I would just be part of the black matter eternally, not as atoms because Einstein wasn't created yet.

Science and Stephen Hawking's theories of many universes aren't proven or real, as all theories. That, for me, is just nothing too important, and stopping there, to say bigfoot could be real. It is the same kindergarten theology, but to dream is everyone's right. It is also to keep for themselves if it has no value. The media needs something to fill open spaces on papers and time on TV. It goes after anyone for helping them earn a living, spreading nonsense like a fire in the corn barn. Why not? A job is a job, but we can separate the garbage from it.

The dark matter can accommodate an infinite number of universes, because it has no borders; as Hawking states that there are multiple universes, his idea of the universe is one surrounded by neighbors. It encounters other universes' frontiers. He then knew the line, but I wished he would go with me on the *Starship Enterprise* to show me the borders. I can take a picture to show to everyone, especially my grandchildren.

He will be remembered for his dreams, as a number-one Hollywood contributor for the fantastic fantasy series and movie pictures in the adventures in space, bringing joy to all as the marvelous, colorful world of galaxies. It even helped Disney, the number-one genius in the entertainment industry. I just wished he didn't have to include the black holes, because it gives nightmares to children and some adults alike. He affirms Satan is real.

This idea just came to my mind now as the fireworks are doing a good job bringing us colorful explosions. I am enjoying a New Year's celebration from my windows in Copacabana. A big fireball shoots up and then explodes in dozens of balls. It keeps exploding, creating smaller ones. It explodes again. People scream, excited in happiness, while I am taking notes. Scientists understood the big bang and the multiple universes. The banging at my windows never stops, because it's the wishes of the Almighty God, the only one who controls the radiance of the dark matter, and for the dreamers, they better keep dreaming, because there are no taxes involved. I wish the big genius would come to my windows and watch the millions of people under my feet, nine steps from the enormous sidewalks, and feel what it is to be a human being. I doubt he will be staring at the monitor eternally, because it will be food for his thoughts.

Hollywood interviewed me twice about my last book, which I was writing on the ill-fated *Concordia, God! The Realities of the Creator* (2010). They are thinking about a movie. I would call it: *Concordia: The Ghost of the Titanic*, and it has been in the planning stage almost two years, because they love the ghosts in it, but as there is no sex involved in the stories, Satan and his family are missing on the decks.

Everything is written because I began writing the book aboard on the thirteenth. It sank on a thirteenth, thirteen months after its publication, but it isn't enough for the sinners. I refuse to add pornography to the story, and the interviewers say there is a public demand for it.

Seen from the air, our planet is blue and white, as are the celestial colors, but it is full of mysteries, due generally to its habitants' incapacity to see the striking white lily in the middle of the huge lifeless lake in the center of Rio de Janeiro, the marvelous city, polluted by godless people. The pollution from open sewers has killed and is killing millions of good-size fish, along with any vegetable life in it, and the stench reaches the heavens and the gutters of hell, as it goes through the whole country. In Brazil, everyone is Catholic, while all the third-world countries pull the strings, blaming God for the natural catastrophes, as his punishment.

The only one not blaming God for anything is the scientist Hawking, because he has deleted anything that begins with the letter G from his screen and yanked the page from his dictionaries with the word *God*. Now no one can call his disabilities a curse or punishment, as stated by him when interviewed on *Discovery Channel*.

When I was nineteen to twenty years old, my family was classified as a high-class poor family. (Brazil has names and ranks for everything, such as "sidewalk residents" for beggars.) Having graduated from junior high, as note 10, I had to answer the call of patriotism, serving one full year in the army (1953), but I spent my time not soldiering or putting bullets into the target. I was good, but I was drawing everything they needed on the paperboards, because today's technology was only a dream, and I am good in straight or curved lines as a freehand.

When I was discharged from military service, I told my mother I was already an adult and ready to confront the world, not waste my time in a university, just to learn one path on the road with 360 directions.

I told her because she was my guardian angel. I had already had fifteen minutes on tape about my radio program on soft music. I was the programmer and speaker. As soon as I had some money to open a small office for advertising, using my drawing and design talent, I offered my work to the general stores and industry. She began crying, and I still don't know if it was happiness or begging her saints to guide my path to success, and she guaranteed I was her baby eternally, but told me to keep both eyes on women and trust no one, but I failed to follow her advice.

My first step was to confront all adversities, already holding tangled in my first investment the famous functional ink pen *Parker 51*, in deep blue and cover in eighteen-karat gold. For me, it was a weapon, which followed me all my life.

I went to the office of the *Estado De Minas Gerais, Syndicate Of Professionals Journalist*. I entered the big office on the second floor of the best building in town with twenty-two stories, I faced eight hard-looking seniors. One of them said sarcastically, "Listen, nice-looking lad; are you in the wrong place? This is the *Professional Journalist Syndicate in Brazil.* Are you delivering anything?"

I looked at them and thought I was finished. Should I give up or fight to the end, as I had nothing to lose? I was just beginning, and I recalled my mother advising me when someone tried to stop my running to push him to the side firmly, or he would come back.

"Sir, I am a professional journalist, as I have a radio program from 11:00 p.m. to midnight every day, called *The Beat of the Night* [it was on the air only two nights] on the best radio station in the state. I was told to classify for my test for my journalist identification, also with my abilities to read and write about many things. I can paint, draw, and design whatever, and my rates in schooling were nine to ten. As for my self-taught English, I rated as better than any professional teacher from the fancy English Cultural Society. I opened an office as my first enterprise in advertising. I can do my own drawings, and I had *Antartica Guarana Brazilian,* the number-one soft drink, as my first client."

They all looked at each other as I had my fingers crossed behind my back and whispered to Jesus to do a miracle. They all retreated to a back room and came back shortly with a piece of paper, a bottle of ink, and a pen.

"Well, you gave us a nice speech. You have the experience but are a very young teenager. Let's see if you can do better in 180 seconds, writing down why you are one of us."

I felt a chill in my bones as they came back to bury me. Having fun, I grabbed the paper, sat down at a table, took my *Parker 51*, and said, "I hope I won't disappoint you all with this pen, the most powerful weapon in the world in the right hand, as said Thomas Paine, the number-one leader on the rights of men, as you all could be my grandfathers. I am not intellectually at your grandchildren's levels, because I didn't come from a golden crib. I earned this pen and my position with lots of sleepless nights hitting books, not watching ball games or in groups at bars or squares."

"Son, your three minutes is running out."

"What do you want me to write about?"

"Convince us you are born to write, and this is you first and last chance, as you now have less than 150 seconds."

I looked at them, as being judged by Satan, and it was me or them in victory. I was the only one with something to gain. I firmly put my pen on paper, and words began exploding onto it. I never stopped to know until my eighties, despite all adversities, but how far now before I am going to stop, or are there any stops. That's the question or mystery!

I still hold my journalist identification, as a small letter folded into two parts, with a black-and-white, well-done picture showing my serious teenage face. I took and still take my writing seriously, because from any writing can result negative or positive thoughts, which can destroy or build souls (see it in the last pages).

I carried it near my heart all my life to remind my soul we must demand our rights, while respecting the rights of others in our path here or there, because we exist in an eternal life, where there is no death. There is a transition to better worlds called evolution. Time never stops, and all the good deeds are ours to keep. As for the journalist ID, once you earn it, no one can take it away.

14 Does God Have Free Will?

Intentionally or not, I sat down in downtown Rio in one of my favorite coffeehouses, as they have the best, most delicious foamy espresso coffees and the patisserie tastes like it was made by angels.

Alongside me were two young pastors, armed with Bibles, like soldiers holding their guns. The younger asked the older, the one with all the wisdom, "Brother Josef, God gave us free will, but why is writing in the 'word' as 'not even a leaf will fall from a tree, without his will,' when there is no free will for him?"

His fellow priest looked at him, as a representative of the high authorities, took a deep breath and making the sign of the cross, humbly said: "Dear brother Joel, you are right, it is at the 'word,' and no one can change that, not even him, because his laws are unchangeable. It is his will, and he knows what he is doing as the creator."

I stared at both as if they had a problem, and both in one voice asked what my religion was.

"All the good of them all, provided it is logical and not theories and fantasy, because I am old enough to have common sense and reason!"

They almost dropped their fine demitasses as I said: "In my free will, I can make my own decisions, as is the right of all of us, including him; otherwise, he would have given it to us, as many of us can choose hell or heaven!"

They simultaneously smiled, holding my hand as a team. They asked where I came from. I answered, "The moment I leave, I will give you my business card." They now held the sacred books with gratitude or as a protection, as if getting ready for questions and answers not found in their religious university teachings.

"Obviously, God has free will; otherwise, he wouldn't have been able to give it to us, as intelligent human beings and his children, or we wouldn't be his family. He can change any of his laws and regulations, but we can't do that. We can only follow him as the Creator.

"The difference between his free will and ours is that, as a Creator, he can change regulations as time keeps ticking on our behaviors. But some things are unchangeable, like whether to be born, or not to be born. That is his decision, as is the death of the carnal body. It is unchangeable, because it is our path to the eternal spiritual world. It is very important; we must harvest what we sow."

"But, Mr. Billy Moreira, he is omnipotent, the almighty, because he came before the universe. It is hard to believe he can change his perfect, unchangeable laws."

"Listen with attention, my lads. His laws are unchangeable as far as what is logical, or the universe would be an anarchy or chaos. We are created, like everything else, to follow his rules. Morals are for us as intelligent beings, as soul or spirits. We are free as he is, but we are the ones those laws are for, because we are the created in our world, in his universe.

"The fall of a leaf without his permission is just an expression of his authority understood by great prophets, spiritualists, and psychics, as Jesus is the great one. He can change as he wishes for our benefit, but he always keeps the ones necessary for our spiritual grandeur.

"We have our free will to choose between good or bad, to learn a little or a lot about culture, but it is deeper than that, because we are enriching our spiritual path to enhanced worlds, and I hope we are giving in writing what is clear in my conscience.

"As we become more profound in spiritualism, science advances, dominating the negative, microscopic world of germs, advancing the medicine and bringing relief to the carnal bodies, and showing us the marvelousness of infinity. Evolution is part of our pathway to the stars.

"The Creator had all in his designs. I bet it took him several millennia, almost a countless number, because the addition or subtraction is infinite, as confirmed, adding another number on the final counting.

"As our intelligence advances, as does everything else, a great scientist, Hubble, dedicated himself to giving us the eyes to infinity, perfecting the telescope so we can stare with our vision and mind very clearly; even in our backyard, the splendor of the creation is there as in infinity. We can be happy, being healthy or not, walking, running, or in a wheelchair, black or white, man or woman. We are all part of it eternally, as our changes continues, because the spiritual life is ours.

"We are the unique planet in the Milk Way constellation, holding life, as far as we know, and the only way to reach beyond it is as spirits, as our voyage is guaranteed after our existence as soul begins here, as we exploded when a sperm runs toward the ovum and makes a soul. Then, we go to the marvelousness of the infinite as a spirit, and this continues the sequence. If there is no understanding in my words, it is not my fault if there is no one to listen, because evolution is a slow path to eternity."

I gave a hand sign to the waitress, because she was around listening and smiling like an angel. She reminded me of Carol, and I asked for three cappuccinos and a small tray of their fine cookies, because I was dying of curiosity about the reactions of those two youngsters. They had just begun to see and feel the diversity of the world. The challenges are constant, and many times we feel it is getting too heavy to carry, but we must survive to the last end. There is our glory. I found out writing this book and feeling the words just flowing onto the paper, like water from an eternal fountain of light.

Those two future pastors embraced their journey in the name of love. The ascent to the glorious path to illuminate the other is a hard one, but the reward will be for them, because all the deeds are counted. I hope they reach their eighties or even beyond, because there is a lot to learn when love is involved.

This time, they got up to embrace me, in grace, for my spiritual philosophy. I was always embraced by the Bible, as the traditional black cover, and they looked at me as seeking a light not found yet, putting me between the sword and the wall.

"Young men, I can give you the logic of our free will; if he didn't give us some of his in our creation, then it wouldn't be a marvelous world for us but mostly like that of hobos or slaves. We would just have to obey orders and no choice at the dinner table, no color, and everything we do thousands of times during the stage called earth.

"It begins now, as I ask you in your cup, white or dark sugar, sweetener or just as it is. Even in the polemic stages, as after a long romance, we get married and shortly things don't work well, and there are the choices of a painful separation, but it is always a choice, being part of the free will."

Free will is the greatest gift to us. It is almost as important as our creation from the composition of matter to an intelligent human being, including disbelieving him and not showing his face.

Well, this scenery is for the ones looking at a great painting, and with dull-witted minds, they say no one did it, it came from the air. Ungracefully for him and not me is Hawking, as the Creator put him in front of a colorful monitor, twenty-four hours a day for fifty years to have enough time to find him. He can wait, but Hawking's time is running out, because the door of the unknowing is near. He said he doesn't want to die (yet).

The Nobel Prize committee should create one for the lost souls, fighting the strong currents upriver, not aware of floating backward into the gusts of hell, created by their negative free will.

Only the ones who "have eyes to see and ears to hear" will notice how wonderful our creation is, in our world and in God's universe, because the ones with long telescopic vision are too high on a sand pedestal, disguised as scientists. The visuals are staring into illusions called theories and put on the site they dream of a senseless fantasia, bringing only "double talk," not helping our necessities on earth, like fuel becoming scarce, famine, warming of the planet, social anarchy, and on and on, because problems or no problems in unreached spaces are not our concern, but the breakdown in our planet right now is, because it is here and not at out of the way places. It is only possible as a spirit by the ones already in the afterlife, because the dreamers are here, like the wolf looking outside the strong fence at the fatty lambs, saying they are too skinny for dinner.

The ones who are born blind or become blind a few years later can smell the flowers, the aromas of food, the warm embrace of human being to human being and so on. The son of musician found great consolation and took it on as a hobby or profession, as the singer Andrea Bocelli, Ray Charles, and an endless line of souls, which glorified and glorifies the Creator, receiving a life with happiness as a bonus.

The ones without the material visuals are always smiling, not looking, but feeling as if they are surrounded by good people, not seeing themselves aging, the filth of third-world countries, where a large population lives like animals in favelas (ghettos), as open sewers divide streets, and while listening to the sound of music, the bad news as part of the media is not on the agenda.

The blind see God and do not demand his presence physically but spiritually. They don't blaspheme or commit suicide.

"Sir," they said (I feel spoiled being called 'sir,' guaranteeing my sureness as a senior, of having crossed all the bridges), "God being above all, does he see the future? If so, how can it be?"

"This is a tough one, but let's use our logic, as everything seems to be mysterious, and because of that, you feel like you are in a labyrinth without exit signs. I will choose a simple way to help. The word *omnipotent* means he is always in control of his creation, the one who began and commands all the shows, being good or he wouldn't have had the intelligence to create marvelousness for us. We are it, because we can appreciate it, and I can talk and listen, dialoguing as intelligent person to intelligent person, about all the concept of existence, and it is absolute evidence of an eternity beyond any imagination. The colors of the innocent smiles of children dominate our stone hearts as he does us indirectly. Yes, God can predict the future as he planned it. He gives us all we need to sow to then harvest, but if we don't do it, then we will starve. This is a prediction and can be or not be a reality, because it's the free will involved. Death is not a prediction, but the future, as it will happen as a sure act."

"Sorry, sir, I feel my question is a sin, but does God have two faces?"

"Well, to have two faces for us, is like being cynical, but if you are analyzing decisions, he must create continually as events turn up in his free will. He can change things around like us as being the owner of a business. He can change our orders to meet the necessity of events, but the ones at the base of success will always be the same."

In my restaurant, I changed menus and arranged the dining rooms, but food must be served to customers as advertised. The same goes for the author while writing, he can change the subjects as often as he writes, but it all must be related as a book. He can cover many subjects, such is religion, science, food recipes, and comments related to gastronomy or a mixed pot, as I love to do.

The great religions, like Catholicism, Islam, Buddhism, followed by many smaller ones, affirm the Creator doesn't change his laws, and it appears logical, but as my example above, it is like his unchangeable laws are the federal ones for us in government, and the others are the local ones made by city halls, not affecting the main ones from the top. It isn't easy to explain in words what is in the heart, but it is the sincerity that counts.

I spent many hours at night with my small Cessna 182 RG, between hell and heaven in the big black in the freezing, tense hours, above the skies of New York. It was worth every minute, because it is like being in a dreamland of real fantasy, where the curvature of earth is on the ocean line. Gravity's power of attraction maintains the water bent perfectly as bamboo tubing, and it is so glorious, as I felt like I was in a Disney cartoon or an imaginary dream. My wings are fake, but I was blessed by the Creator to know. All I wanted was to be closer to him, as a good son seeking his father. He may be busy in his designing of the deep glow of the dark matter, but he is never out of contact.

After that, I couldn't dare to look at a monitor, seeking celestial understanding, as it is now just a fake reflection of the real picture. It is like tasting an imitation of lobster meat and comparing it to the real one. It is beyond my common sense to hear someone being snooped on by society dare to say not privately but at the four corners of our suffering planet that there is not a Creator behind the majestic show of beauty and power.

I believe this one person is paralyzed from his feet up to his head. He is just avenging his pain, dragging with him as many souls as he can to echo his cry of misunderstanding, as a mortally wounded gladiator in the arena facing all the thumbs-downs. No one is ready to die, not with a blade at his neck, facing the guillotine drums. The best way is to go take a nap and be awakened by a beautiful angel saying, "Pappy, sorry, but I asked to come first to be able to help you and all my relatives and friends or anyone on our new journey into the spiritual world, our eternal fantastic world, where there is no more death, and our family won't be separated because God is love."

When flying, I could feel the freedom of just listening to the gentle roar of the 230 horsepower, whirling the heavy propeller, seeking the path to dreamland, as everyone dreams, but few dare to seek it. I could feel I wasn't on the ground or in the water but in the open space as our own imagination. I felt pity for the ones with big mouths, as they can't feel the freedom from gravity, where infinity is the limit.

Having us as his image, he gave us two hands with five fingers on each one and feet with ten toes, two eyes and ears, and most of all intelligence; otherwise, we would be rational, not landing on concrete, asphalt, or a grass runway but in a cooking pot.

We gave him headaches because of our free will. We began wars, like Hawking's black holes, scaring all the living people, including aliens. They are also human, merciless beings like an animal pack on its prey or a hungry shark biting and yanking flesh for their meals, such as when bombers hit cities. Innocent people, not being with their families, wait in horror for death, as God forgot to include morals in his drawing.

I refuse to believe he gave humans the knowledge of atoms and the control of them because we are predestined to assassinate his children or even to destroy earth for good, terminating our planet, because those devilish souls are the rotten apples in the barrel. Atoms are for the industry a cause for energy in the form of electricity, but the ones using them wrongly are going off track. The effect begins here in the vulnerable carnal body. Infections are armed so no atomic blast can reach it, because it is deep in our veins, arteries, skin, and even inside our brains, saying we must love each other.

"Carol, come here, please. Could we have three espresso coffees please and this time, a half dozen of those soft-crunch chestnut cookies?"

She smiled and said to keep calling her Carol. I say it felt like she was an angel, as if her name were Raquel. Suddenly, I heard a voice whimpering, saying, "I am here, Pappy."

I am amazed, feeling the lack of common sense and reason from human beings in general. They have no logic at all, and I am sorry to say it, but I feel it would be irrational for me to hide my feelings. Many times, I look at a huge crowd, and I have the sensation of staring at hobos, as programmed, intelligent individuals, not with senses to see spiritually and not only materially our beautiful world. They wouldn't scream at traffic, because the teenage driver at the back had his radio blaring loudly, and a gentleman at the front stopped and went back with a baseball bat to smash the windshield of the new car. The fearful young man stayed helpless, and if I hadn't jumped in to firmly calm the insane driver, there would have been another fatality.

I gave both my personal card, and the next day both called me to thank me for my intervention. The ruthless one paid for the damages, while the youngster also learned his lesson and did not press charges, as I asked him not to. He never put his radio above normal sound, because both learned the hard way.

To choose in free will to be good or bad is our decision, and the second choice has dividends to pay. We have the right to use this gift to love each other and be Good Samaritans, as everyone benefits. Tomorrow it could be our turn on the road. The dividend to pay is called by religious people sins. The payment begins here, and they could follow eternally, as for this reason, many "sinners" run away from church.

My two young evangelist students were staring at me like I came physically from a faraway constellation or just materialized, and the youngster asked me if I was a pastor from "the old school."

I did get up, saying I had to go to the bathroom. I passed "Carol" and paid the whole bill, giving her the cards for my new friends. I exited through the back door. I told her that they asked for me, she didn't notice the senior with white hair and a short beard. The bill was on the house, as a gesture of good will for evangelical students, for the business cards she found on the table.

I went back a few days later to that coffeehouse, and "Carol" told me about the two future godly men. After twenty minutes, they went to the bathroom and looked everywhere. They asked her for my whereabouts, and she confirmed what I told her. Both had to sit down, shaken to the roots. They said, "Sister, Jehovah is merciful, because we had an encounter with a celestial angel, and he gave us enlightenment from heaven."

Getting home to Copacabana Beach, I had two messages on the land-line and one on the cell phone, saying they were counting the minutes until they saw me again, if it is God's will, and it made my day.

15 Who's at Fault If Someone Doesn't Believe in God?

We know the author by his book, the chef by the quality of his gourmet dishes, the medical doctor by his reputation, and the saint for his miracles, the pilot for his accident-free records, and so on, as say the British: "The proof of the pudding is in the eating."

Our existence in a body, as we begin our "body and soul," starts not in infinity, but here, among a man and a woman; that's why the Vatican began the *celestial concept of marriage.* According to history, the Roman Catholic Church, as a powerful institution in Europe, at the ceremony to bestow God's grace in 1563 at the Council of Trent, put the nail firmly on the sacrament of marriage as a common law, and the whole world adopted it, giving morality to an immoral society, guaranteeing women some part of men's macho world, not just as an instrument or as a slave or a tool, guaranteeing the heritage of the family.

As we notice everything is based on solid rock, the concept of a Supreme Intelligence, what some fear when doing wrong, is a signal in the conscience that something is wrong. Others do it with respect and gratitude for his creation. This same society is composed of mainly good souls, while the minority of the wrong ones try hard to bring anarchy to the family. The famous "black sheep," disguised many times as a scientist, says there isn't a Creator, because the telescope doesn't reflect one but infinitely colorful galaxies exploding and continually expanding as new, perfect universes, while he is chained in the incongruous human body as a consequence of his disrespectful challenge to the Almighty Creator and affronting his creation.

I had almost two hundred employees at my steak and fish house in Miami (about thirty years ago), and suddenly, everything was going wrong from the kitchen to the dining room tables. I stayed the whole day analyzing who was the devil responsible for the instability. It was a waitress and her niece as the second-ranking chef in the kitchen. I had no choice but to tell them to go back to the employment agency for good, and as they left, peace came as a blessing.

Imagine a planet without God; it would be complete anarchy, as there would be no hope or faith and no religion, because the rock foundation would be sand, like in the dark ages. Courtrooms would have no defense to judge our planet, because the laws are based on God's morality.

Well, if someone, blindly in his conscience, feels like a godless citizen and if he tells some of his relatives and friends, they are not obligated to listen but can tell him to keep to himself. I have seen innumerous cases during my eighty years. Sooner or later, everything would go sour in their lives, as health or financial disasters and aging faster than normal. They would see the Creator, and then peace would begin its process in healing.

Recently, I threw a party (sixty persons). Someone, after two glasses of wine, began to try to tell all the happy ones that God as a creator makes no sense, but I saved the night, telling him to cross the avenue, tell it to the fishes, and not return anymore. I opened the front door, and peace took over.

But a well-educated man, having a very serious illness that keeps him mummified after fifty years, not to enjoy life at all at its minimum as a normal human being, his is a "brilliant mind," qualified as a genius. I agreed, because fascists like Hitler, convince nations to build the largest atheistic or godless army to share a vengeful single consciousness, which stems from the fascist's nonhuman condition.

If it spreads as a mind epidemic of devilishness, the first to vanish will be *marriage as the creation of a family, as the Catholic Church* so arduous protects morally. Lately, they worry this concept is being lost. The values of a godless world are bringing us to the dark ages, the divorce rate is out of control, and men and women just live in promiscuity. There is no God to answer to.

The idea of a godless existence is a no-man's-land or better a world where anarchy can take over, because there is no conscience to bother with. There was no marriage concept before 1563, as there was no concept of family. Society was a pack of irrationals conveniently for men, who were physically strong. Women were their servants, because there was no Supreme Intelligence, as everything came from an emptiness.

It is absurd that now our planet has worked so hard through moral laws, a pack of godless men and women have a free media. President Obama represents the number-one country on earth, the citizens are hard workers. *In God We Trust* is on the medal to honor the number-one godless scientist Stephen William Hawking, who gave while bitterly sitting on a *divine nail.* He promotes openly that we are godless.

Hope is not a consolation to the godless conscience, as everyone like him shouldn't be paralyzed, but as we have God. In him we trust, and to be paralyzed for life, first, we must give a reason to the Almighty to earn this life incarceration, as a reminder that life is like playing with fireworks. If you do not respect its power, anytime, sooner or later, it will blow up in your face.

He has no logical right to use the media freely to help him spread his hopeless idea and theories of a non-God. In a country without a president, anarchy would then take over to announce a "brilliant mind" full of theory. As for me and millions, it is just mythology and not fact. Just telling about dreams and fantasy doesn't help society or put bread on the family table or save the planet from ecological tragedy or being incinerated by atomic blasts.

As I know and affirm, warning often comes from angels, saints, prophets, geniuses or no geniuses, a sign in the sky, a prodigious child, or an eighty-year-old senior, who in the past twelve years began writing books based on spirituality and logic and the foundation of our progressive intelligence called common sense. He progressed in culture and collecting the material to explode a moral big bang, advising it is not too late while *In God We Trust.*

Science and religion don't go deep into this area, because there is nothing to reassert, but they keep one eye on infinity, not realizing the distance reflected in a deep powerful lens is an illusion as it will be reachable maybe by all of us only as spirits, which never come back to a conference on earth, as not being obligated, they wish not to come back.

At a famous bar, known as the perfect "waterhole," philosophers of all levels showed up as the stress took over. They had famous sayings, and one of them was: "A well-balanced spirit will never go back to the cemetery, because it is just a warehouse of bones." It comes back to earth only for a mission or a punishment. No one wants to die, but no one wants to return, especially the ones sitting down on a "divine nail." Most of the time, they are even famous, because a "nail is a nail," and any time spent on it is an eternity.

Don't analyze my logic as harsh or comical, but a mixture of both, as realities, taking it that charisma will solve the problem better. I never did and will not make fun of anyone, as I don't like it done to me when serious issues are involved. But writing without a trace of essence is like a master creating a gourmet dinner and forgetting to put in a touch of garlic, or worse yet, a dash of salt.

To make my day, it came to mind, back years ago, I had the pleasure of personally going to the great shows of comedy, with stars such as Bob Hope, Red Fox, George Burns, and the Brazilian Chico Anizio at Carnegie Hall in New York City and in Las Vegas. But they are all now history, living in our memories as great souls and now spirits.

In Brazil, as in the United States, I watch TV shows by great interviewers, all in their eighties or their nineties, and they all joke about being old, as a great stage of creation, a path on a terrain no one can avoid. They talk about being on it in absolute comfort, as time comes gracefully.

Salomon: Two Points is a TV interview show, with the Brazilian Larry King. He is near to entering his nineties, but on his daily program, he interviews big names, speaks six languages well, and is even now greater as he ages graciously. He shows life is better when we are at peace with the Creator, because his blessing goes beyond the aging.

Here, as I say, do not wait for the neurons to begin crumbling. Keep busy reading and writing continually, and you will fool the aging process, staying somehow healthier and looking younger than you should. It's working, as I am envied by youngsters.

When someone asks me if I am William Moreira, the old journalist or the author of a specific book or the designer of great sandals, as I draw the feminine feet first, I look around as if looking for an old man all wrinkled and curved by aging and say, "No, I am not, because he's young in body and soul."

I had very busy restaurants in New York and then in North Miami Beach. My contact with customers from all levels of society was for me a great cultural experience. I had boundless acquaintances with wonderful people from the movie industry, singers, chefs of several nationalities, and restaurant owners, as we exchanged visits and ideas, and the simple people, whom I greeted as just as special. They all became regulars, and I felt life in its greatness, because all my life was in contact with people and not facing reflections from lenses or plastic monitors that have no soul.

This contact was and is a great incentive to continue in the business. It is hard work, because we give all our efforts to the public and they give back equally. We need each other eternally, because death on earth is just the beginning as is being born, like in *2001: A Space Odyssey.*

From Miami, one of the best otolaryngologists in the country, *Dr. Joseph Friedman*, came in for dinner. He was a senior about sixty-five, and it was around 1975. (I looked up his name today, but now there are a hundred names, like with mine in Brazil, and which is the right one, God knows.) He is eternally in my heart. A customer told me about his reputation, and I wisely approached his table armed with a nice tray of hot, fresh, crisp potato pancakes from my recipe, just-made applesauce, and the best sour cream. When he saw it, as a good Jewish man, he conquered me instantly with his smile as he invited me to sit down.

He began a conversation, saying he drove one hour for my food as it was highly recommended by several of his patients. By the long line, he knew I had the best food and reasonable prices. After a few minutes, I told him I had had a bad incident inside the left inner ear while deep sea diving, and I lost hearing in it and heard a sound. Tinnitus took over, like the devil, and several doctors from the United States and Brazil couldn't do anything. He told me to go the next day to his office and only the Creator would decide if a successful surgery was possible. I told him next time to use the kitchen back door to avoid the line.

I was impressed with all the letters from patients filling the walls, and the waiting room had no empty seats. With no delay, he examined my ear canal, and looking at the X-rays, he said, "The doesn't look great, but I will do a surgery with God guiding my hand as I seek a miracle. Want to go for it?"

"Doctor, can we go now? My faith is total!"

"My secretary will call the hospital near downtown, Cedar of Lebanon, and anytime there is an opening, as I believe there will be in only a few days, we'll go for it. Son, we believe there is a Creator. If I can't restore the hearing, at least the devilish sound will go away, God willing."

I got up, kissing his hand, and saying his corner dinner table, number 63, would always be reserved for him but with flesh flowers, crisp potato pancakes, and fresh applesauce.

On the surgery table, I was given deep sedation and local anesthetic, because he wanted me to tell him as he worked inside the ear, if the ringing (tinnitus) stopped. It was an incredible experience. I heard celestial sounds, and after a half hour, he said if in one minute the tinnitus sound didn't stop, there was nothing else he could do. Suddenly, I yelled, "God, the sound is over! Doctor, the tinnitus is gone!"

I kept repeating the same sentence, but they held my mouth, saying it was time for me to go to sleep, because I had received a miracle. As I awakened, he was there, smiling like an angel disguised as a medical doctor, performing miracles for those with faith who deserved it.

Angels are all over, especially in the medical and religious fields, as the devilish ones sometimes come as scientist to scare the hell out of us and bring chaos, but as always, they carry a heavy cross or sit on a "nail" until they give up and run to "angel school."

Until his death a few years later, he never missed table 63 at 8:00 p.m. on Friday nights, but I keep it for him with flowers and a picture of him savoring my potato pancakes, as only God could do better (the four-step recipe is in the back pages, and any older child can do it).

The next day, after the miraculous surgery, when no one was around at the hospital, I went down on my knees, and for a few minutes, I cried to our Creator. I did not sit on a "nail," but listened to the birds outside the open window, feeling how great it was to be healthy, to appreciate his universe, our home.

Outside, driving home, I had my windows open to listen to the wind, the horn blowing, planes flying, and even the tires making a wonderful sound smoothing the road while driving north on 95. Like the blind man who got his vision back, the sound made the complement of the visual and sound a perfect world of enjoyment. I noticed the spiritual world and exchanged the brutal sound of hell for normal hearing as a bonus.

A ten-year-old child, as his father, the president of a great nation, asked him why some adults in conversations say God doesn't exist, because nature is all and no one as a decision maker is involved.

"Daddy, can you prove to me God does exist?"

His noble father felt how deep the question went in such a young mind with common sense. He deserved an answer that could help this bright beginner on the long and challenging road of learning. He has a lot of questions and no answers.

The mysteries of life are just mysteries because there is no equation for them. It begins with the word *God* and finishes with God. God is God, or God is not God, because everything is related to the existence. There is no way to pick at its window to look at the brewing pot, because it is guarded in a manner, blocking any soul and then spirit to view or feel it brewing.

His answer for his child whom he knew was above average in school could not alter his way of facing life on earth, as most of us are on a rocky road without exit, having or not a pedestal to lean on. Aging begins at the moment of birth.

He went to the library holding the boy's hand, and there he picked a selection of CDs from the National Geographic and Discovery Channel about the beauty of nature on earth to the galaxies. For two hours, both enjoyed the color of a marvelous display of a fantasy, as our world in God's universe. The father asked the son who was the mind, the Supreme Intelligence, and the Creator of everything, not letting go of any minimal details, as a whole. The boy entranced his daddy: "Pappy, anyone who has a head with brains in it knows there is a God who arranged it all for us to enjoy our existence eternally in a colorful world, with so many different things that we must live perpetually to appreciate it all."

A day before, aboard the cruise ship, I tuned into the Discovery Channel. Children, who are now adults, were born with special physical deficiencies, having arms and legs with only about twenty-inch stubs. Now, with websites and e-mail, they meet and have contact, knowing each other and sharing their experiences. Most of all, they know they are not alone and are doing all they can to enjoy life as happily as anyone else. They are grateful for being born and having trust in the spiritual world where there are no deficiencies.

They go out into the world doing everything, laughing, meeting others, as normal as they can be. They make speeches, open businesses, create businesses, and don't depend on anyone to clean their noses.

About twenty years ago, I flew from San Diego to Tijuana in a rented Cessna Cardinal. I had lunch at an elegant restaurant. I noticed a beautiful, elegant young lady in her thirties at a corner table. The waiters and many people gave her lots of attention, and she always smiled. When she paid the bill and starting walking toward the exit door, she looked at me and said, *"Buenas tarde, Gringo,"* and left, moving like a *goddess*.

A few minutes later, I left. She was by a car talking on her cell phone, and I approached to ask the way to the private airport and also to take a closer look at such a beauty. As I got near, I become adrenaline-charged as I noticed she had no arms. Her shoulders were rounded smooth as her blouse had no sleeves. She was a divine angel. I felt she was a spirit who came there to teach all of us a lesson in humility. She did everything using her beautiful feet, raising them up to her lips, putting on lipstick and then putting it in the side bag. She took the car key and opened the car door.

I tried not to let her feel I had noticed her handicap, and she behaved so naturally, because she was at peace with herself and the creator. Because of that, she was walking, driving, and loving people as they loved her. I asked her if she was married, because I was looking for a beautiful Mexican girl, and she said she was engaged to marry a great doctor in Acapulco, but as a nice man, she knew women would flock to my feet. Then she drove away, living and glorifying our Almighty for the lessons.

How could she have a normal life? I see able-bodied, healthy people not even able to peel a banana or wash a dish, and worse yet, get up in a restaurant and push in the chair, leaving it in everyone's way. They throw the paper at the basket and miss it, but they never pick it up, feeling they are above even angels. They abuse their providers, hugging themselves to death as they reach a deadly weight, while complaining the world is unfair to them. Millions in third-world countries work, earning a miserable pay, the public hospital has no doctors or even aspirin.

Now, I landed in Brazil, in fabulous Rio de Janeiro. I tell everyone it needs American management. Tender tasty beef, fresh seafood, and sour cream is not plain yogurt floating on lime juice. It needs discipline more than ever, as well as free, clean public bathrooms and clean sidewalks without potholes.

The populations on the planet are embracing each other more than ever, as the weather is warming, provoking all kinds of hazards. In Brazil, the south is drowning, and in the north, twelve million people don't have enough water. Their stock died; the only source from the ground became dust. Millions of children have no schools, because of competition with India, while President Dilma is the busiest one on the planet. The people paid for Boeing, publically visiting every country, as she went to the Vatican to be the first one to kiss the new pope's ring, saying proudly to him in front the international media, "The pope is Argentinean, but God is Brazilian."

She didn't think that because God is Brazilian, he belongs to every soul and spirit that ever existed. If she had a little common sense, as a lady from a first-world country, Britain, she would have known the status of American dollars and coins, even before she was born: "In God We Trust."

As an effect without a cause, I ask myself what the next step for me is, as I was firmly ensconced in Miami, and suddenly, I am even thinking of staying here in Rio, writing books and long letters to cynical politicians, confronting all kinds of odds. Some friends tell me to sit down and relax, as my time has arrived. They ignore time, because it is the present.

They advised me to go to Buzios, which is three hours northeast of Rio. It is a resort people dream of, but after one full day, I was on the bus back to Rio, bored to death. Having no challenges, on way back, my thoughts went to Hawking, as being tied motionless to a chair, not carrying a heavy cross, but nailed to it.

His purposelessness and empty existence come from one of the most terrible illnesses. He has not even the privilege of succumbing to it, because to "die is not to die, but to keep going as free." He wasn't allowed to. It came to my mind, as I mentioned in this book, that it was a test for him to leave his black holes in peace and dig deep into a cure or relief for his condition, as help would come, but he wasted it, seeking the faraway hell to rest his case, while denying publically that there is a God, and he got burned.

If he had contributed to mankind, for example in medicine, he would now be on a rock pedestal and not a dusty one, and the world would be at his funeral, saying his name alongside the Creator and not alongside the *king of the darkness's*, the only one at the cemetery as the digger goes home.

When calamities happen, now more often and getting worse, his name is glorified as begging continually for piety, but when the bombers drop their lethal death loads and the bayonet crosses each one's heart, Hawking's God is real. The assassin, proudly in uniform, never tells his victims: "I am killing you in the name of God!" or "God forgive me; I don't know who this person is, but I am killing him like a roach, because my government says he is my enemy, and as such, he needs to be eliminated. It is just a job."

The average human being does make mistakes or errors, Pontius Pilate did when he washed his hands symbolically, or better yet, diabolically, as if cleansing Jesus's blood from his conscience, when sending him to the cross, but when he died, his soul landed straight in the spiritual hell, or the Celestial Department of Corrections for Repented Souls, as the CDCRS, the purgatory created by the Vatican. It's not eternal, just like Brazilian prisons or penitentiaries. The lethal cutthroat goes in at the front door and exits a few months later out the back door, as he had purged his crimes, by putting his hand on the Bible. The Brazilian system accepts it conveniently, because taxpayers save money, while crime and anarchy dominate the country, and it gets worse as long as misery stays around, brewing for *The French Revolution, the Sequence.*

All my thoughts and opinions today have an origin in my getting close to a century of researching in all fields, talking to lords and beggars. The evolution is from science to medicine, as I was born at the right moment, seeing the evolution from propeller to jet and then to rocket under one umbrella. As a pilot, I saw the compass transform into digital controlled by the GPS (global position system), becoming a reality in a few years of piloting my Cessna, and I knew technology was born as God's gift and more is to come. Wrongness will be paid automatically, not as punishment, but a reprehension for us to be humble as a family.

Interestingly, as not to say tragically, people say their wrongdoings were the fault of malignant spirits or even by *Satan and his family.* They affected their behavior, dominating their minds. If that was so, then it would be a free will, but in today's society, the law doesn't fall for that, as religions now give *Satan* a day off. The Roman Catholic Church calls them spirits, as you and I will be soon.

Until a few years ago, in several European countries, when a crime was the devil's fault, the guilty one ended up on the end of a rope, but in today's moral evolution, he or she will get life in prison until their last day, but it includes open access to the library of holy books.

I imagine, as is my right and your right, I do express it, or there would be no rights or better yet, moral evolution. So far, we don't use disrespectful words and expression, immorality and lies, to bring chaos to society and confront men and God's laws, such as in not admitting there is a Creator. This takes hope from faith, as the faithless spread our lives are only what is reflected in our eyes. There are individuals without consciences (the sense of wrong in one's conduct or motives, impelling one toward wrong action to follow the dictates of their conscience).

Those individuals—kings or beggars, atheists or godless ones—not admitting the mysteries or, better yet, the complexities of our incapacity to understand life as it is, where the marvelousness is the intelligence of a Supreme Intelligence, being it to master perfection, here and beyond.

The questions were answered by my collection of private short or long interviews, as time passed in my long life, among the general public, without audiences and not wasting time, especially the ones there when the moment was ripe, in the streets, at their jobs, in my business, at restaurants, in bars, at the scenes of accidents, at parties, during funerals, in churches of all kinds, and on long train trips, such as from New York to Miami and California to Paris and to Istanbul, and Istanbul to Athens. I have taken hundreds of long-distance flights and cruises. I have been in hospitals and prisons. I filled my notebooks to the last iota, but then I stored it in my mind, where is no way of losing it. The important thing is to keep it in the memory bank, as the spirit is perpetual and we can reach it when it is necessary, or we just go back to recalling the data, reclaiming from the past what is necessary, as I am doing now.

Any subjects or controversies generally end with God, the creation of all existence, as before the universe, mythos, and realities. What is beyond the "valley of death," demonstrates that humanity has priority in another existence and not only politics. It does not matter who is famous and rich, but continually, there will be an eternal life at the death of the carnal body. There is a logical reality, even for an atheist. It gives a justification to amend the statement that there are no unbelievers. It is a way of expressing their dislike for our world.

As I took notes, very few ever thought of life on earth as a paradise. The love that Jesus gave his life to guarantee is a life beyond death. It was mentioned by religions and unreligious souls, who said it indirectly, as I asked them to mention Jesus's name. The answer came, saying there is already normality in pronouncing his name, but it is more than that.

In Rio de Janeiro, New York, Paris, Rome, Cairo, Tokyo, Beijing, and Buenos Aires or anywhere, millions of dogs are being treasured more than children, and there is nothing wrong with that, but if for every animal, the owner would adopt a child living alone on the streets of a third-world country in the ghettos (favelas) without care, the earth would be warless. No one would be shot or knifed as assaults continue in the name of miseries.

16 February 5, 1996, the Weight of the Cross and the Ghost's Consolation!

"Pappy, I finally bought the video cassette important to our life. The movie that hit in our hearts, *Ghost* that affirms death is just exchanging the material carnal body for the spiritual one."

Carol screamed with cheerfulness, not knowing her days on earth were counted, but mercy was available in the form of a movie's pictures, touching millions of hearts in every faith.

"Tonight, after dinner together, we will sit in the TV room to view *Ghost*, which came from heaven as a logical consolation, as death is just leaving our carnal body to the eternal fluid, spiritual one, staying sometimes here for a while, as necessary. I can hardly wait to see it again!"

"But, Carol, my angel, how many times did you see it?"

"At least four times at friends' houses, but I need to show you, because it could come in handy one day. Death can take anyone by surprise, and this movie can help souls as spirits."

After dinner, she dragged me and her eight-year-old son, and she had baked a creamy strawberry cheesecake from one of my recipes, but hers tasted as no one else could do it. It became famous as "Carol's angel cheesecake."

Ghost hit full blast the bull's-eye of people's hearts. These people have been wounded by the enigma of death; as consolation, only Hollywood could master it. They did not miss the target giving comfort to adults and children alike, *Ghost* was put on a high pedestal as hope in a solid base of rocks.

"Carol, who's the one affirming this fact as spiritual. Inclusively, we are the ghosts going around scaring to death the mortals."

"Daddy, I think Hollywood only mastered it because it gives

consolation to anyone when death strikes. We all became lost in our grief, even knowing we will all go through it, but it is good to know as we educate ourselves. It is insurance for bad moments, such as sickness. Special accidents can happen to anyone!"

My wife was visiting her sick mother in her country, as I had one eye on the screen and the other on Carol, feeling something was going to happen. She smiled many times, but her face was one of deep concern. She was getting closer to me, while she had her son on her lap.

When I finished the movie, her big eyes looked in mine, as if trying to tell me something important. I always could feel something wrong was going to happen, as I said to her, "My dear baby, please tell me, are you sensing your time with us is getting short? Do you have a serious medical problem? If you do, I promise to keep the secret as long you want!"

She stayed there holding my hands for an eternal minute. Her face was down. I could see a drop of a pearl rolling down her face, and then she said, "Pappy, sometimes we feel like we had a bad dream, as if death will take us away. This presentiment has been haunting me for months, and even now, I drive under the speed limit, always on the right, and paying attention to any details that could be lethal. I am concerned I could become paranoid. As I go to work in Times Square, I go for a few minutes daily to a beautiful Catholic Church and say to God that if possible, I want to see my son become an adult. I want to go to his marriage, but, Pappy, if something happens to me, think about what *Ghost* is teaching us, as no one vanishes with death. When there is love involved, we will meet again."

One week later, on the ice-cold morning of February 5, 1996, it was windy and the ice-covered ground defied gravity. She was thirty-three years old, when, as she crossed on the green pedestrian light, a huge truck from the *Times* newspaper went through the light. The driver alleged the brakes had frozen and he couldn't stop, but it was proved false, and he then admitted running the light and not seeing her.

This affirms that when we must go, nothing will stop it. It was a confirmation of karma, destiny, and a mission, a trio as real as it can be, for the ones seeking spirituality and not black holes for moral comfort.

She died instantly as a young beautiful lady and not as a pile of wrinkles as an old woman, maybe alone somewhere, waiting for relief called death or paralyzed in a wheelchair as a mummy, seeing and feeling a wrongness in godless world, where there is not even a hope of an afterlife. In this philosophy, some of us find a light in the tunnel of bitterness, because believing in a godless existence is like digging our own grave or spending a lifetime confined to a wheelchair having no hope of walking.

At the funeral, more than eight thousand people lined up in front of the white casket wrapped in light pink. On the top, I had a three-by-three-foot color portrait of her face at the age of twenty, and another when she was six. Everyone wanted a copy of it, as I had painted it. I had a picture of the painting, and I mailed it to everyone, as two wallet-size photos. It took me three months of work, but the photos were sent to all her admiring friends that attended her funeral. More than two thousand replied saying that now she would be waiting for us. Meanwhile, she was there to help, as they know what her name is now: "Carol, my angel."

Now, imagine if I had Hawking as my idol. I would have been hopeless, suffering eternally as a godless man, hating the driver, a father of four small children, attracting all the devils, living a miserable life, as Hawking does.

Our home in Weehawken, New Jersey, near Boulevard East, became like a tourist place, as for almost three months, people from everywhere rang our bell with a single red, pink, or white rose or lily and embraced us. They would say that angels come and go, but they come back to comfort us when a tragedy is involved, and she now is one of them. At the main office at Marriott Marquis Hotel, in the heart of Manhattan, we and Christopher, her beautiful eight-year-old charismatic son met for coffee and cakes, as thousands of her job colleagues wanted to meet the son and parents of an employee who became the angel of the eighth floor, the reception, restaurant, and coffeehouse area, where millions of guests went by, and she, with her manners and appearance, were engulfed then with her smile and charismatic way. The general manager said that without Carol, it was like the lights went off, and it took months for the staff and many guests to go back to a normal life, as she is always mentioned.

I went back last November (2012) as seventeen years rolled into history, and only a senior waitress remembers Carol, as the wind whispers around the stones at the cemetery. There is no one to listen or read the names and dates engraved, saying history is the past. Now is the present, and let's keep moving because time doesn't stop. It heals wounded hearts.

My wife had *Ghost* for many years as a consolation message from Carol, and when someone had their heart fragmented, she would invite even a small group for tea and a slice of "Carol's strawberry marbled cheesecake," but I did the baking. She played *Ghost* on the big screen, and reality or not, everyone felt consolation was for the ones staying behind, not faithless theories, with a solid affirmation of *Satan* disguised with the door of the dark hell, as a black hole.

I sometimes watched Stephen Hawking on the monitor, and Carol used to tell me to change the site, because he needed prayers, as he was a lonely soul, stuck in his own body, unable to enjoy even a minimum of life.

I asked her what the reason was, and she used to say he was being punished for his tongue spreading a hopeless existence, and it isn't love and imagining. As she used to say, if the whole planet believed him, then we would all feel like we were in hell.

"Pappy, we must have compassion on the ignorant ones; they do suffer because of it, as believers of a marvelous existence without a Creator and organizer, but I doubt it isn't true. I feel sorry for his condition, as it isn't God's fault, but he is in playing with fire."

The unknowing death, even to the ones fanatically holding to religion, believe heaven is waiting for them. It holds hope, even to a floating wine cork, saving them from drowning.

Religion is a necessity to the ones seeking answers from mysteries beyond death. They feel disheartened to know from science, through the incredible telescope or other apparatuses, getting near celestial corps in the infinite space, sending data that earth is the only planet in the solar system able to sustain human life. Our descendants will not live, because it has only 24,859 miles at the equator, as a jetliner will go around in forty hours, at 621 miles per hour. About 71 percent is salt water; we are practically waterless, because 2.5 percent is freshwater, and in the third world countries, it is mostly polluted by sewers and industry, as 70 percent of it is used for agriculture and would be food on our plates.

Logically, this book is not meant for the "brilliant minds" or "geniuses" because when earth gets mortally wounded, they will jump into the spaceship *Enterprise* with their families and vanish to hell—I mean, to the next celestial body a billion light-years away from us, land there, and when the aliens with three eyes and fish skin open the door of the spaceship, the powdered dust from their bones will be the witness of a lost civilization.

I hope they understand that continuous human life is right here, and for the ones already with their souls prepared, like me or anyone above their sixties, it wouldn't make difference as our time is up, but we think of our children and grandchildren as they will die a slow, painful death, and our worries for them come as a package, being intelligent human beings with a God behind and above all. Our hope is things will be fixed.

I asked Carol when she was in her teens what would happen to us if earth were to be destroyed by a meteor or atomic blast, and she laughed and said: "Nothing special, even an arrow from a rascally little devil will blow it up like a balloon. We will float in space like astronauts, waiting for the next big bang, but without Hawking's black holes."

I consider it almost impossible for an author to describe the pain of tragic death of an angel, if he or she doesn't feel him- or herself deep in their hearts, the bitterness of drinking this chalice, as a blind man describes what is light or color. I hope I did it in color, not black and white.

It would be a fantasy and not reality to imagine what it is to feel a needle inside an infection, and worse yet in the spine, for an examination to see if it is bloody. When I was twenty-five in New York City and someone hit my car, I landed in the hospital.

I can describe piloting an aircraft with severe problems while flying on burning adrenaline, in a passion to survive, because I was one of them, and I was the only one to see the violent death of his daughter as I did. I will feel the deepness in my soul, as part of being on earth as an intelligent human being, as everyone tries to believe it only happens to the guy next door, as a defense, but it isn't guaranteed. When the phone call comes at 4:00 a.m., we wake up screaming: "God! Help me!"

If it is just imagination, then it is a gift of consolation. Besides the contacts of a child with spirits or devils, which scared me out of my wits, presenting as monster or angels. It was as a blessing, because fifteen days after Carol's death, I heard loudly and clearly her voice early in the morning, "Pappy, read page fifty!"

Then I heard her voice twice guarantee it wasn't a dream or a nightmare. This short sentence changed my way of facing my life and I began writing realities, trying to help the ones in need. I began seeking to understand the pains and greatness of life, as we are here in a transition as a tourist, because our base home is beyond death.

As my heart was fragmented and nothing at all could heal it, then came the balm from him, as the charisma of Carol's voice as my sister. He is our father.

After six hours of seeking the page fifty of several books, I opened the right one. It gave me logic about "why life is like this on earth." It was in my Chrysler minivan, right in the front seat in a box where I kept books. That particular page was marked as waiting for me.

I will not ever tell the name of the blessed book, so as to avoid religious and even scientific inquisition. Now I want to wish for eternal peace for everyone, because at the negotiation table there is no bloodbath, but reasoning, because patience comes when the fruit becomes ripe in time, when it gets to the right stage of taste, for our savoring.

After the cascades of tears of bitterness stopped, a few years later in my apartment in Oakland Park, Florida, as I was watching the most beautiful sunset ever, I heard her voice again: "Pappy, life is beautiful; it doesn't matter what happens, because we are God's children."

I grabbed my *Parker 51*, and again, the words began exploding on the paper as the sunset rays illuminated my thoughts. There was then born a consolation book, dictated by an angel, using her death experience to help us.

17 Not All "Brilliant Minds" Are Worthy of the Title

I was the first one at the buffet. I sat down and began my first line today. I continued my writing of this book, and as I put the pen on the paper, I just wrote Johannes Kepler (1630).

It is interesting, if I can say so, how our memory bank functions, gathering information even before our thoughts get involved. Now, I had to dig deep into the name of the one who first affirmed to have understood the mysteries of the orbits of the celestial bodies, as they maintain their elliptic trajectories (paths in curvaceous line). There wasn't much time to help him as deeply as today with our technology. The chicken's world ended at the fence. Progress and evolution move slowly in time, but they never stop, as adding it all through the logic of those "brilliant minds." We can reach perfection, but for the ones based in the Creator, the solutions came on a silver platter.

Theories that do not become realities aren't any help, only dreams and fantasies. They are only profitable for Hollywood in entertaining the public and for their amusement. The same goes for religious leaders, as science holds no solutions in theories. There are no solutions dragging centuries to millennia, like one-note music. The only successful composition is Ravel's *Bolero*, but it stops when we pull the plug.

In colonial Brazil, the *Bandeirantes* (country flag holders), carrying the Brazilian flag, went west like the pioneers from the East (New York) to the West (California) discovering the country in wagons. In Brazil, they walked because of the dense vegetation and irregular terrain, and everything was carried on mules' backs. After about two years away from any contact with civilization, on the arduous mission to expand the country, they discovered a huge snowcapped mountain range (the Andes), like the Rockies in the United States.

If (the number-one word in the dictionary, in describing theories), they had gone over it, the Pacific Ocean would have been there waiting for them and they could have had a day on the beach, but in ignorance and tired to their wits, they returned to Rio de Janeiro, and Brazil lost the coast-to-coast privilege like one of the greatest countries on the planet, our United States did, especially after adopting *"In God We Trust."*

Isaac Newton (1727) was a genius, or better yet, a brilliant mind, because he went beyond theories. He ran after realities and not just dreams and fantasies; it goes well in the same pot with mysticism.

He did not sit down day and night dreaming, seeking black holes to engulf not only the universe but our only salvation from the grave. He was seeking the only word meaning the eternal afterlife: hope. It is backed up by the icon of eternity called God.

Only browsing his biography, we can feel how he gave us some general findings, not to scare us, but to bring solutions. He did not seek dreams and fantasies, because dreams and fantasies come from a playback of our conscience, as an anarchy of senseless displays, nightmares. We all should forget it and face the day normally. At night, a little prayer would help us. So far, we had done nothing wrong to disturb our subconscious.

As we study and analyze the lives of those "superbright minds," which brought us advances in disciplines in the past, without all today's advanced technology, we must appreciate that they opened doors for the future, bringing goodness to our difficult life on earth.

Newton had a tough early life and did not enter a universe or school on a red carpet provided by rich parents. He worked his way up with his ability. If he had been glorified as Hawking is today, but on God's pedestal, then I would stand up to applaud.

As a true monotheist, Newton saw one God as the masterful Creator, whose existence could not be denied in the face of the grandeur of all the creation, as just common sense. He was backed up by Einstein centuries later and denied eternally today by Hawking, and his last book *The Grand Designer* is just an illogical book. It was his last hope of being the son of a Creator and receiving some alleviation from his ordeal.

Going back to history, we analyze the lives of those great souls, bringing to us progress, as evolution, giving their colleagues through the centuries and millennia the path of continuity, as seeding for the future generations to live and enjoy.

In 1687, Isaac Newton published his single great work the *Philosophiae Naturalis Principia Mathematica* (Mathematical Principles of Natural Philosophy), as he showed how a universal force, gravity (as an apple fell onto his head) applied to all subjects in all parts of the universe in addition to his studies or calculations of the infinitesimal calculus and the new theory of light and colors. Hawking, like many others, is navigating on the wagon of wisdom, from those such as Hubble and Newton, to enlarge the theological world.

Book Description: *The Grand Designer*, release date: February 21, 2012, ISBN-10:0553384666, #1 *New York Times* Best Seller

> *"When and how did the Universe begin? Why are we here? What is the nature of reality? Is the apparent 'grand designer' of our universe evidence of a benevolent creator who set things in motion—or does science offer another explanation? In this* startling *and* lavishly *illustrated book, Stephen Hawking and Leonard Miodnow present the most recent scientific thinking about these and other avoiding mysteries of the universe, in nontechnical language... brilliant and simple."*

When I read the above statement by the most serious newspaper on the planet, I was stunned by how this article passed the logic screening test fit to print on the *Times*. It was more that of a *tabloid* (a nonsense paper, as "bull crap," a popular British expression (BS) widely used anywhere).

Newton used his common sense on the level of his intellectuality as he advanced science without borders, never disputing his grandeur in the direction of a better, more understandable universe. He was unlike Hawking, who elevated himself to a high IQ, qualifying openly as having a genius mind, while stone in his material body, about which he had no choice, as he is created and not the Creator. He should have spent his carnal time researching for solutions and getting relief from his "crucified body." Some people believe, and an older priest affirmed years ago in Southern Florida, that Hawking is the reincarnation of *Judas Iscariot*, and this time, he is betraying God. He landed on his cross but stayed on it a lifetime, so as to regenerate his soul, and as always, my *Parker 51* kept taking notes.

Newton, as he stayed in the shade of an apple tree, did some calculations when a ripe fruit fell on his head. He looked up and wondered why the apple did not hit the bird on the head. As an intelligent man, he was dedicated to being the number-one scientist of his time. Gravity was born, and he began as a "brilliant mind." He was supported by the Creator, not staring at a monitor, but in the universe and glow in the dark matter, not seeking a creator. He was predestinated in a great mission, and as he accepted it for his glory, he then avoided the wheelchair's mummification.

Newton spent his life intellectually and physically as a great man, in a vast field of science, ahead of his time, preparing the way for others to continue on his path. He put everything into his work, admiring the Creator and not questioning how existence began. He used what was at hand and understood it best to enjoy it.

He noticed the group of the illogical, the short, visual minds gossiping in the corners, and even denying the master of it all, afraid of facing realities, as time rolls into a spiritual dimension we belong to, while the illogical circumscribed in illogicality dies as soul and spirit.

The studies of Newton are considered the height of humanity, and in stature, he was tall, athletic, and nice-looking, adorning squares as a reference to a mind that gave so much, not expecting society to kiss his feet. His glory was internally in peace with the Creator and not to be cynically called a "brilliant mind," while betrothed in a wheelchair. He did not want to be a "nothing," a zero appearance in immobility, as a reflection of his mind. He bragged his IQ shot up to the top 250, as a showoff. It is his excuse for not having a Creator of the majestic creation, but he can't even get up to do a minimum or even know his own voice.

When it was time for Isaac Newton to depart to his dimension, as we all will do—it's guaranteed—his funeral was like that of a king. He deserved a pompous one, and general society came to pay their respects and thank the one who gave so much graciously. He kept hope as part of our eternity and because of that he had it. We build or destroy our own futures.

He never dreamed of an absent existence; it wouldn't be worth a dream, as even to seek comfort or to look at a meaningless life as an irrational. He did not dream that a threatening meteor would eliminate life or that the existence altogether of the universe would be destroyed by a black hole, because his time was very precious, not for bragging, but helping and enjoying life in the first-class, as Europe is good at that.

He sure didn't have to sit down on the chair, stressed as could be. The one who qualified himself as having a mind with an IQ passing over 250, or better having an infinite number of neurons. Anyone can state that, but the acts or deeds are the ones who will separate the Newtons from the Hawkings.

Everything in the universe came from the big bang, as an explosion, and its fragments went in every direction in the infinite glowing dark matter. We know, even just looking at it at night, feeling an eternal aquarium as having all the celestial bodies. That's when we stop, because it is the demarcation between our limits of intelligence.

If we try to go over this limit, death will end the carnal body and there is no exception, because the pain is infinite, as infinite as the ignorance.

Gravity is invisible, as is almost everything else, at least to the human eye, being the reason the word *theory* is widely used. To see is to believe. It is now myth, because millions of things happen around us with today's technology. It is felt as electric energy, shown in the gauge, as almost infinite communication signs surrenders us. Here and in infinity it is felt as real because it works. Why not for the spiritual world have God in the centerpiece?

Flying my little plane, I had or have my earphones as the contact to the ground or for plane-to-plane communication, as it could be anywhere, loud and clear, giving us moral feelings that we are not alone, and they also hear us as alive and well.

There are no lines to drag, but a solid invisible one and visible perfect signals. The GPS moves on the monitor, like it is alive, giving us perfect directions, digitally, as voiced, in a choice of many languages. Driving in Greece, I froze to hear the GPS talking in Greek, but with the press of a button, it said, "Good evening, welcome to GPS international," but the street names were in Greek. I just followed the lines and numbers.

It goes on in infinity; as Newton found out, there is an attractive force controlling the perfect movements of all the celestial bodies, and it came from somewhere. It is as far as we can go. If I would call a scientist a "brilliant mind," I would call Isaac Newton number one on the list, like everyone else.

The problem today in qualifying people as saints, "brilliant minds," or geniuses, is they hand out diplomas, as it is, with the title "doctor" in third-world countries, and if you have money or are well dressed, everyone will say, "Please, Doctor William Moreira, here are the cards for your suite," or "Paging Doctor William Moreira, please, sir, contact the information desk; the captain is inviting you for dinner."

When the Vatican began their innovation, facing today's world, they started on the right foot with the Vatican science's new approach and invited Hawking or he invited himself to test God's people or them him. Being a doctor south of the border means also being in science for a long list of "brilliant minds." As far as I know, he kept his mouth shut, or he would be out of orbit, because the pope is serious-minded, not playing Russian roulette with our Creator; otherwise, the Church wouldn't survive.

Lavoisier (1794) is on the list of "brilliant minds," as the father of modern chemistry. He did not demonstrate theories; otherwise, there would be fantasies, as the godless ones have no support for their theories, because even the ones born blind can feel his presence. As "nothing is lost, but transformed," based on his confirmation, that includes the soul. When he was devilishly guillotined, his spirit left the carnal body, and he floated away in his glory, because he didn't tell the world how intelligent he was or brag and say how brainless others were. Despite his cruel death, it was instant. Of course, the main suffering was the sensation of it. His life and death had an eternal future in the spiritual world, because he admitted the Creator was the master of everything. His death was quick, but during the waiting moments, I imagine what it would have been like if not for faith in our God. Hawking feels his time is getting closer; he is already crooning, as he isn't ready, because he does not see a continuation. He is only responsible for his actions.

The matter that composes the incalculable vast eternity, before any big bang, in how it appeared, formed, and is being formed came from a source, and if you become paranoid, take a cruise to nowhere, as I did, and let the overpopulated ambience distract you as a prevention to craziness. (I hope I am hitting the bull's-eye. I have mentioned many things often, because not everyone finishes reading a book in a few hours.)

Colossal or microscopic are the invisible gears moving continually, not by a man-triggered force, but gravity. It affects all the creation, perfectly rotating, colossal as any mass, and it is beyond our imagination how it works. To discover gravity, it took just one apple to activate a chain of events, but the source is unknowable; unknowing is our beginning of our creation as human beings from a microscopic zygote. Is it enough ammunition to satisfy our quench for a Creator?

I repeat it again: if wasn't for our intelligence, there would be no universe, because we, as a pile of atoms, were as good as a rock or a chicken. If there were not a God to be blamed for it, then Hawking wouldn't be famous as a godless man. But my question is: *"How come a genius with a brilliant mind, having, as he states, an IQ of 250, and having been gifted by nature, doesn't accept a Creator? Otherwise, the Creator will compete with him. Didn't he dig deeply to find his way out of his heavy cross?"*

He also rebuked Jesus and the cross, as the cross means the utmost suffering. His ordeal is as real as God and Jesus. One is a brilliant mind, while stepping on our toes. I, William Moreira, never need any evaluation. Once, one of my employees called me a jackass in front of everybody. I just walked away and came back in five minutes with his last check. He thought I wouldn't survive without him.

Hawking stepped on more than God's toes. In our hearts, disregarding compassion for our suffering as the only ointment for the wounds, the hope of a Creator, his last paycheck came from heaven as body mummification, unless someone puts down his head and says, "Lord, I am an unwise soul, and now I realize it and beg for your forgiveness, as I have learned what love is. Now, I hope to be able to help myself do good for others."

One word can destroy any relation, and four can heal it. I will put in the bank account one hundred dollars for whoever can mail me the right sentence that works for anyone; it is very simple.

The reputation of science not believing in a divine power of creation is because of Stephen Hawking and his followers, but we can't condemn a bag of apples because one or two are rotten. Some have stress, others a fatal sickness like Hawking's, and others have tragically lost family members, like I did my daughter Carol; that would be a reason to go against the Creator, blinded against any marvelousness at our feet and seeing only the negativities.

The ones in desperation accept it as karma, destiny, or part of their mission on earth. They have a better understanding about death than the ones revolting against the creation, as there is no way to change it. Without comprehension, nothing could be done but bow to the force, as the condemned go to a long jail sentence and know rebellion against the system is no solution.

I meet many people in the same situation as Hawking. I also was paralyzed from the head down for almost thirty days, and I hung onto God's neck for the entire month. I ended up at Saint Mary's Hospital in Hoboken, New Jersey, after falling down the nine steps in my house on an icy day. One morning, I woke up and just naturally began moving slowly. It was a miracle. About two hours later, I was normal, but God heard me at least one million times on those days, lying down and moving my arms and head while the rest of my body was frozen and black and blue. I had to be in every stage to write this book. It isn't a fiction, but the realities are as raw as they can be.

The doctors believed my nervous system was in shock, hardening almost all the muscles, but I chose to stick to God. It works all the time. I never blamed the Creator at any moment; on the contrary, I confessed that if I did wrong and was paying for something or if there was something I should have done, I was sorry for my ignorance and would wait as long as necessary.

How can we go against a creation? Things happen not only to John and Mary, but to all of us indiscriminately, as we all comes from a zygote, grow, and create a family like anyone else, and then we age and die. The majority of earthlings, such as you and I, accept it, and if not, the suffering is just suffering, like the atheist's carnal life. For them, there is no afterlife, but as anything else, they have no choice. For the lucky ones, this is the Creator's laws, changing our monotonous lives to busy ones as his continuous creation.

Those knowing all and knowing nothing are the renegades, to a point, in unmerciful claims to be more intelligent than others. As some people sometimes say, I am very intelligent, but I take it as idiocy and walk away. Imagine when Hawking opened his hobo voice announcing to our surprise that his IQ is between 230 and 250, the highest for a "brilliant man." As there is no effect without a cause, he is my hero in this book for his Nobel Prize of blaspheming the Creator all his lifetime, as he complains of never being mentioned for the prize.

In the last ten years, as I have interviewed many people while taking notes, I asked about the "brilliant mind," and eight out of ten said Hawking is a dreamer, as is anyone else, while his quantum theories are not saving the planet. He condemned earth as it would reach its one thousandth birthday, but why is he not offering the healing for its survival, as he found the cure for his illness?

All the geniuses in science are giving Hollywood a lot of fuel to glorify the cosmos with fantasies, and it is great, but Carl Sagan (1996) was in cosmology, astronomy, astrophysics, and so on and the genius Walt Disney was to the entertainment industry. His TV programs were in everyone's homes, as we felt we were being transported to the grandeurs of life and his voice, as Carol said, was that of an angel bringing good news. Infinity is a real glory of the cosmos, and we belong to it.

He made people understand that science is like anything else. It is necessary and unites the distant religion as scientists can be anyone. He explained to people that science is not a difficult thing to understand and made the government look at science as a necessity for technological advancement.

Sagan had so many rewards that he could fill the cosmos, NASA became like home to him. He left us twenty great books that anyone could read and more than six hundred scientific writings. His sagas will stay here forever, and we all miss him.

Hawking did not do anything to help earth stay healthy, because as he said, "Earth will not last one thousand years, and we need to find another place."

I wonder if he is leaving us a fleet of spaceships and a planet to land on, commoditized for all of the billions of us, and that will then be a genius deed. It will not be theological, because I hear BS every minute.

Carl Sagan left us a legacy of beauty and hope, as we visualize earth and beyond. People stayed home to watch his great presentations, and his name is now part of the landscape, as the life he had in bringing evolution beyond the borders of space. He spoke about black holes but not as an enemy of earthlings or the creator. Prisms seen from different positions have highlights that differ from other points. The same goes for our different ways of seeing the world.

Some "brilliant minds" could just be "obfuscated" to others, but the opinion of the majority is what counts.

As I mentioned before, you and I have the right to protest or appraise someone, even not having the support by the majority! I do it without any regret, because logic and common sense, which is responsible for our evolution, should be counted as IQ. A 250-point IQ did just pass a kindergarten test, as when it was given to me, after a few answers, I found it so ridiculous that I ripped the paper test, saying intelligence is known as the author by his books.

Logic or common sense don't come free. They are not bought at a counter or acquired with a diploma at a university, but by walking through the narrow door of life, leaving the dark tunnel of ignorance, following the light of intelligence. Decades before, we didn't as brothers, because it wasn't time.

To have a high IQ is not to fill books with theories or write a million quotes, because everyone loves to write quotes and theories. We have so many of them that they would make a line all the way up to the last galaxy, and no one was stricken by the one who came, living for a while, but staying to eternity. The ones who live by the sword, dies by the sword.

After an eighty year journey, I didn't succumb, fortified with my cosmos pen, the *Parker 51*, as it opened my "narrow doors of ignorance," which before that I couldn't reach.

I had found solutions, which are illogical, such as to say the glowing dark matter originated from nothingness; there was no universe to make a mockery of science and fools of the population.

I came to the conclusion that as time rolls, our evolution materially doesn't match morality. It goes up and down like a Russian mountain or a roller coaster.

As I aged, our planet came to a big leap in technology, science, and medicine, but not really in social behavior. Human beings are not being programmed like a hobo. In our free will, we can make aberrations, as generation from father to son, don't match social values from a recent past.

Just about thirty years ago, if a senior boarded the bus, subway, and so on, immediately a younger person would get up, but today, all over the planet, practically no one would offer their seats. In Brazil today, in every transportation, there are seats orange in color with big signs all over the walls, including at the stations. Every five minutes it is announced they are for seniors, and if one is not available, a younger man or woman must offer the senior the seat, as it's the law.

But I and others must open my mouth loudly to demand our divine rights, before we fall on the floor.

The same goes in all aspects of life beginning with music, as the great, big bands and their soul-soothing music are now pieces in a museum. Taking their place are groups of cave people, yelling while jumping like apes, in Rio's City Hall. They spend millions of dollars yearly on events in front of my windows in Copacabana Beach, as culture for the younger generation, especially the ones from the United States and England. At night, even with soundproof windows, no one can enjoy an evening at home. The same negative phenomenon is all over Brazil as on this planet. Going to a CD store, my only choice is the classical music or the radio programs.

One of the first things I did when I went overseas was to visit the art museums. (I was a regular at the Metropolitan Museum of Art in New York.) I avoided modern art like the devil running from the cross. My thirteen days visiting the Soviet Union involved a tour of opera theaters, and classical art museums, including the Hermitage. To see my desire for fine realities and whether I "have or have not talent," look at the last pages of this book for my designs. I would first draw the feminine foot as reality and then put a shoe or sandal on it. Today's famous designers ask me why I have to do the feet first, and here is my answer: "Because I know how to draw and do not want to put imperfect lines on paper. It would be like hitting the drums and calling it great music or a primitive way of communication."

Now, in today's society, the majority does not follow the good of the great past. Remarkable were the ones who gave greatness to our society, calling them prodigies and now distorted as "brilliant minds," "geniuses," or "IQ gods." Are we in the Roman Empire, where citizens worshipped their emperors or generals, as someone with a theology, as a great one, or faced being crucified. Today, it is classified as jeering by the intellectual ones, as a moral decadency, despite the grandeur of technology; it can also destroy, using atoms to make the awesome *atomic bomb*, which can be our end at any minute. It isn't the privilege of the two big fish anymore, but all the fish in the aquarium. India and Pakistan have beggars on the sidewalk, but their arsenals can destroy each other and have all the others countries as the appetizer. Fallout and radiation have no borders.

The moral on earth is reflected in the countries' leaders, as China and Japan fight for a rock in the ocean and now the North Korean military—a half dozen brainless ones with their finger-puppet—menace the world, as if they have an atomic bomb to extinguish South Korea and their country, and only rats would be left. Because of a few, millions will suffer, and because of one anti-God, the planet can be destroyed. Because of a divine punishment, Satan will be on the loose.

Many foreigners tell me the United States likes to play cop for the planet, and I tell them thank God for that, or they would be speaking in German or Japanese, as slaves, and today terrorists would be the boss of the world, as everyone is the target. Keeping order on our sinful planet costs billions of dollars and lots of American blood, while everyone else sits around the arena, pointing fingers and doing nothing, like barking dogs.

There is no moral to respect seniors among the young ones of today. How are the IQ tests to know who's more intelligent than others? And those getting on this pedestal are the ones suffering in their paranoia.

I read about paranoia and all the studies on the disturbances of the ones going around trying to be above others, using today's different logical society, where moral values are not a priority. They are contemptuous of their negative physiological lives and call attention away from their problems and their inferiority complexes. They feel they are low in IQ or are being confined by a physical handicap, as Hawking, and their sourness goes first to the one known as our Creator. He is not sitting in the corner giving handouts to those rebellious ones, so they state, "God doesn't exist," knowing for sure he won't show up to defend himself. The effect can be felt, as those people are sitting on the "nail," while Stephen

Hawking is at the front of the line pulling the rope (see "The Nail and God's Dog").

As evolution keeps ticking with time, the progress in one main religion is remarkable. It is now at the top of list and is now reelecting its new leader. Those representing 132 countries of different religions came in, demonstrating the fraternity in our polemic planet, which is based on one Creator. Jesus Christ came here on a mission of love. Imagine if Hawking would show up with a placard, saying: *"I represent the atheists, or better yet the anti-God; there is no Creator, as everything comes by itself. There is no need for a God, according to my calculations as the number-one 'brilliant mind.' My average IQ is even above 250."*

Meanwhile, he is sitting down on his "nail," and the moment he is able to ask to go the bathroom, it takes too long, but the diapers are around, as the odor takes its action.

The wind is nothing more than the movement of the oxygen we need to live, as anyone knows. It can be felt but not seen. It causes things to move around, from sailboats to our hair. Even the irrationals sense it, staying airborne all day. When piloting, I checked the windsock for direction, as it helped the aircraft to lift off, demonstrating very clearly the ones looking and feeling in our "to be," as there is no word to humanly describe our universe created by a Master Supreme Intelligence. The wind is around, but it is visually ambiguous. We are not yet prepared to visualize his grandeur but must continuously progress in our evolution in the free will to one day have the contact with our celestial wind.

Pilots feel the Creator as we are above the city lights. We can almost touch the dark matter as it glows. We can just accelerate to deep inside, but not even the ones who went to the moon felt they were being in it and then looked at earth and saw the grandeur of it. They were just staring and grasping as they floated in the dark matter, like fruit on an invisible tree. Interestingly, the Universe-Mother existed prior to anything else, as we originated from it and spilled on the big bang as our eternity.

Happiness is to sit at a table, whether in a palace or in a ghetto, being healthy with family and friends or in a public place enjoying normal human life. It is to be at peace with nature and not be bragging about being a genius, while sitting on a "divine nail."

In 1991, while in Boston, Massachusetts, I took my four grandchildren to the marvelous New England Aquarium, and we had a great day. Outside, the weather was as cold could be. The giant glass cylindrical tank contains two hundred thousand gallons of salt water, and six hundred species swim continually in one direction, from the left to the right. It reminded me of the universe. The irrationals followed the attraction of gravity, including the huge turtles and sharks, the galaxies, and the smallest planets, and here and there, one would dash against the circulation like meteors.

As I told the youngsters my makeup, they loved it and considered their grandpa a genius. That's why when someone comes with theories, those smart ones would be classified as "brilliants minds"; those grandchildren are so innumerable as to make someone's fan club or atheist club functional.

Later, my oldest granddaughter told me she believed more than gravity was circulating the fishes like celestial bodies, because behind the gravity is an Intelligent Mind, the one who created it, who thought of having a pulling power to keep everything together, but he keeps his eye on it. She was only fourteen, I asked her where she heard about that, and tapping me on the back, she replied: "Common sense!" Then she continued, "The magnetic force was noticed by Newton, as gravity. It has everything under control, like a Swiss clock with its moving parts. The biggest corpus attracts the smallest, and everything is programmed with such precision according to the fluctuation of mass, not defying the natural existence of body or the Creator's command. What is doing all this continuous creation up in infinity? Only a manic mind would say there is no intelligence beyond all this. He is only attracting attention or is plain stupid!"

If there was no intelligence, like the Creator, as part of the dark matter, the seeds would grow, because when the signal commands it is rising because of the temperature and the season controls it. Everything else would be any life, as we come also by seed where our DNA is complete, including the intelligence, karma, destiny and mission. It is proof of a Creator who is watching us with free will.

"Heidi, I am proud of you, you are a 'brilliant mind.' I give you a top IQ of 250!"

"Grandpa, I would rather be a humble girl than be crucified by my arrogance, because to live appreciating what is offered is everything, like now being alongside you having a great day. Also, the IQ test is so dumb, because there are questions and quick answers. A streetwise person can do it, while someone with a ten in studies would fail it, or a Henry Ford or even Einstein would fail, because they don't have to impress anyone like you, Grandpa."

Now, she is going through college, works hard in a great job, and lives in New Jersey, happily married. We communicate often as she continues being herself and not a false personality, suffering the consequences.

If it wasn't done by a Creator, where did the seeds of the Universe-Mother and the universe or universes some from, as said a bright fourteen-year-old girl, and I confirm the salmon would jump upriver to lay their eggs. Everything would be stagnated, as the sun wouldn't show up every morning, or I would not have been born after a sperm dashed to the ovum, as my carnal father wouldn't have done his part like the salmon. I wouldn't be here as a rational person, writing to you. Neither would you be reading. Let us be humble enough to love each other and to deserve an eternal existence, because everything is as real as it can be. It is not "to be born or not to be born." It is not a choice, but a blessing. The effect of a cause is the most correct and divine law, because if we do not go around stepping on anyone else's toes, our toes won't be stepped on.

18 I Can Prove There Is a God, through All the Splendor

In January 1994, thirty-nine years after the death of Albert Einstein, I was floating near Princeton University (New Jersey) in my Cessna Cardinal RJ. I requested permission to land, as the landing strip was in their backyard. I just walked in through the gate of this renowned campus, as my mind thought about Einstein being there for his last years on earth.

As always, I was curious to know; as I stepped onto their grounds, I thought how I was born in a third-world country (though the Brazilian government brags they are not) but not mentality. I came to New York City at the age of twenty-one as a professional journalist and felt at home. I stayed there the rest of my adult life, admiring continually the human perfection that results when culture is part of it and the desire to be ahead in evolution. It is demonstrated in the impeccable landscape. The reason I mention it is because many Americans blast this great country, ignorant of how living in some other countries is like living in an underworld.

When I go to visit south of the border or "across the pond," I don't stay only at the tourist highlights or take a taxi and see the peripheries, and I don't fly and take the bus to other cities. A few days ago, I went on the cruise ship anchored in downtown Buenos Aires. You can see it floating in open sewer water from the emasculated downtown just few miles away. The landscape is like an underworld. You would feel sorry for the majority of the lower-class citizens, such as your waiter, the bus drivers, and so on. They live every day as public assistance is at ground-zero level.

If you would see the world as I do, every day, opening your eyes, you would thank our Creator to be in the United States of America, because you know better.

As I walked in, I felt like an invited senior professor, as the students passing me smiled and saluted. I felt like a human angel just dropped from the sky, a fake one, because my wings are aluminum feathers.

I walked around appreciating the privileges of being there, when a younger student told me I should stay for lunch, because it was delicious. The direction to the cafeteria was to follow the aroma.

As a chef, I noticed how great and well-presented the buffet was, and with my tray, I walked around and found three senior professors. They pointed at the empty chair at the table. I hit the jackpot, because one of them was old enough to have met Einstein, who had lived in the area since 1935 and died there. He wasn't a professor at the university, but he went there often on other business.

The older one was going to retire soon, and he had spoken many times with Einstein and the great scientist used a comment when people asked why he didn't believe in God. For him, it was a great distress, as he never did say anything with respect to this irrationality, as our own existence is all we need of his presence.

At that time, the professor was a young man, a student and admirer of Einstein and his ideas. Being charismatic, he felt the great scientist set his humor side. Once in a while, he had a few minutes of chatting. He noticed we were all rational and created to have contact as humans, having more capacity to rationalize than others, to feel how united we are spiritually. He knew Einstein felt lonely when he became a widower, but his death in April of 1955 gave him all the freedom he needed at age seventy-six.

Many times, I have said to myself that many things that happen to us are not a mere coincidence or luck. I winged seeking a lunch spot, and as Einstein came to my thoughts, I found the right nest to land in and met the right soul to take away the negative fable of a great man related to our Creator. I needed it. It came at the right time, planned ahead as this book.

I had to say good-bye to the wonderful professor. I usually said so long, but at his age, I doubted I would meet him in our world again. Meanwhile, a tempest was brewing overhead, and Caldwell, my home base, was twenty dangerous minutes away. I could never forget him; he was as gracious as someone could be, sharing his intellectual experience and growing more food for my thoughts. Many good souls come into our world to share and give good news and not anarchy, as in religion or science. When my wings were ready to roll over the concrete, I shouted: "Forgive me, Mr. Einstein, with your knowledge and wisdom and your marvelous difficult road on earth. In your experience, can you give us the proof of the existence of God as a spiritual being and the creator of everything that exists. Otherwise, nothing could exist!"

Then the sound of the first thunder came, and the propeller furiously swirled its 3,000 rotations per minute and pulled me forward into the unknown.

The first cause is for us to understand the effect of things. We have to observe things. It is impossible in nature for something to be a cause without an effect or an effect without a cause. The effect is a cause in the theory of the big bang, where everything began as a universe. It is the effect and not the cause, as it came from the uterus of the center glow of the dark matter, as the Mother-Universe.

As I began leaving the wet ground and watching the sky, I felt the ceiling getting closer and a black blanket like a black hole ready to eat me and the plane, like in a nightmare, but I was confident that Caldwell was only fifteen minutes away for a touchdown. However, I knew hell could break loose in just one minute. As I kept the acceleration at its top red line, holding it to a forbidden zone for a few moments as I counted the heavy raindrops like bullets. I could breathe as I skimmed the rooftops of those impeccable, historical little American towns, and the wheels touched down as it skidded on firm but wet pavement. Taxiing to the hangar, I pulled back the acceleration and pushed at the brakes; I closed my eyes and praised our Creator for the flying angels and sending us Einstein's legacies, thereby illuminating our paths and making our journey beautiful, though not easy. Suddenly, the call from the ground came, asking if I was all right from a windy and wet but good landing, and I came back to our world.

God never was or could be created, or he wouldn't then be the Creator. If we continue seeking him, it would be eternal, as the universe in any of the 360 degrees of direction wouldn't have an end, but we can intelligently stop at the borders of the next universe, before trying to keep going to an endless road and onto another universe, which could end in the glowing dark matter, as I call the eternal voyage.

One day, humanity will realize we create nothing but only put matter into matter, building or composing any of our inventions, taking the material from the natural surroundings. We also come from the material of the environment as flesh and bones, like the irrationals, and no one can stop, though they foolishly try to slow, the aging or decaying process. Time rolls continuously, counting down individually for each one of us to reach the zero point of a new beginning. As said one of the greatest philosophers Saint Francis, "Dying is that we can live again eternally, as it is a natural process and we are part of nature, obeying intelligently all the laws of it, as being souls."

We have at hand all the ingredients to make and create a child, bread, a house, a car, a seed to harvest, a weapon to destroy, medicine to alleviate pains, a fluid matter to dream, and the list goes around the first galaxy and back. There wouldn't be any phenomena or miracles, and Lavoisier guarantees because that as everything else, nothing is lost but transformed. I always speak of the dead ones as alive, because the carnal is incinerated or buried but it is involved continually in nature, as from it flows the energy from its brain as a spirit, which appeared like the big bang and continues its journey. It is not lost like fluids; it doesn't get cremated, as an energy that pulses by itself.

When we sow thousands of pounds of seed, we harvest billions of pounds of grain. A farm could begin with few incubated eggs and have millions of chickens. A scientist researching could put his skills to developing possible compositions from nature and necessities to make our society better in material comfort. Religions can do the same for our moral hope, but no one is creating anything at all, now or ever, because we have the flour, the yeast, the water, and the heat to bake it, and the necessity to eat for survival is already programmed.

As we are created and programmed with all negatives and positives to choose from with our free will, the Almighty, the Supreme Intelligence, the *big designer is taking notes* on each one of us on his divine computer, as his job. He is staying out of the spotlight; that is the reason many of us are ignorant and nested on a pedestal. I am trying to contest those who feel superior, forgetting the composition of their bodies qualifies them as beginners, as being descended from the monkeys, because a few hours after we die, the body instantly becomes just a composition of a pile of putrefied flesh. This is truly an expression for all those above their IQ to know if there is no spirit, then we are nothing, just the whisper of the wind.

We create nothing, as I said before. We put the created with another created and bingo we have a hobo or a son or daughter. We have gears or a chain of wheels and energy by water pressure or steam (locomotives not long ago, such as the beauty Big Boy (1935)), and then came diesel power products, such as gasoline, the movement of engines up to today, and so on. Atoms, solar energy, and wind energy are here as we are, from a source orchestrated by an eternal fountain.

The whole universe uses the endless power of attraction as discovered in the logic of Isaac Newton. It is denominated as gravity and is a pulling energy controlling all the existence of celestial bodies and much more, including its tracks. As a continuous source of energy, it is supernatural for our capacity. We do not understand anything at all about it, as to its origin and on, besides from an Intelligent Spirit, as the source of our existence. I do understand the complexity of creation. As we are scratching it, we should just accept our marvelous existence and then be happy, and that's final.

"To be or not to be" became famous, as any Brazilian street sweeper or worker knows it, but lately, I began putting on to paper such expressions as the following came: "To be born or not to born, it's a mystery," "To attract or not to attract is the energy of gravity," "To suffer or not to suffer is our choice," "To love or not to love is in our heart," as we all should live as one family to bring whatever is necessary to those alongside us on the same journey, and "One as all, or all as one." Such sayings as "He ain't heavy, Father; he's my brother" stayed in my mind when I read it in 1975, at Boys Town, Omaha, Nebraska.

Every saint's day, as I began writing as a young journalist in Brazil, I heard and hear too often many people saying, as if giving me good advice, that I should not get involved in matters that could bring me problems from someone else—politics, religion, science, morality, individualities, and so on—but as I found out, if the average journalist and writer just followed the regular flow, then the progress on earth would stop.

If I feared the retaliation from the negative ones and I didn't take my views seriously, I would continue hitting the same keyboard and then going through my door beyond life after death, and later, I would be sorry for not doing my work on earth, as it would then be too late, because the best teacher of them all stays behind, as physical and moral thing called "pain and incertitude," and all of us, sooner or later will claim his name: oh, God!

The most interesting things are gravity and relativities, words used often when trying to impress the ones off orbit as humans. It is not the rotation of planets or their orbiting paths. It is more than amazing; it is tilting, which any unschooled farmer knows as common sense, having timed when to do the seeding for the right harvest. The Mayan Civilization knew about winter, spring, summer, and fall, as their tomatoes would grow.

The precision of gravity is perfect, but where does the gravity come from. It just angles the planets, as it, like all the others, pulls, running against the tide! Please, if you consult the genius Hawking, he would look to his monitor for another theory, but remember my e-mail has a foot-thick lead wall impenetrable to theories. More and more, I read on the sites—lately, I am seeking the logic of the illogic on the marvelous Internet, as it is just at the tip of the fingers, with a letter-size to fit any retina.

Calculations of theories to satisfy men's egos now fill thousands and thousands of pages. They are not facts and evidence but at least give food for thought, as a demonstration that we are rational, but not all demonstrate that we all are yet at this level, as the ones proudly boast there is no Creator, making a mockery of themselves, challenging the majority to an irrational level. Maybe they should watch *Ghost* as an entertainment. It could trigger a relativeness in their neuron systems, and there will surge a black hole that will vanish eternally the way they focus on life's creation that beyond all the conception is a mind ahead of theirs, as life itself.

A sense of humor marks human beings as intelligent, because not only do our lamentations make us extraordinary, but the humorous faces of our souls do as well, because a good joke can trigger laughs, changing moods momentarily. There is a beauty in our existence, as a smile opens doors and wrinkles a face, as mine did after eighty long years as a student at the perfect college called earth.

If we pay attention to the wind murmuring while we are in the plains, looking at a breathtaking sunset, it will be bringing a message as it did to me, while vacationing (1964) at Cape Cod Bay, in Provincetown, with my wife, son, and Carol as a two-year-old baby (today, she would be fifty-one). While holding her in my arms, I stepped up to a tall grass sandbank and stared at God's postcard as it reflected from heaven to the ocean:

"Imagine! You are incredulous; you're intellectually visually blind if your spiritual one is not connected to it, confronting such a natural splendor as an irrational, not wondering who is the painter beyond it. As a Supreme Intelligence, as the Master of all existence, he is here even before time. It is inexplicable and mysterious, now and ever, but what counts is you belong to it, not temporarily, but perpetually, because you can think and wonder where he is!

In front of your carnal eyes, the array of colors provides scenery not even the best cameras are able to truly transmit to your mind through your own retina, but you can visualize mentally as in the afterlife. It then can be truly as perfect as it can be.

While your feet are in the swamps of hell, created by the challenges of the carnal life, just remember there is a reason for it, as there is no effect without a cause. Ignorance is not to know, and it gives unconditional opportunities for you to upgrade your evolution, always toward a better world, but whether to delay or upgrade is up to you. I gave you free will, because I created you as I did your mother and father, as my instruments of love.

Transmit love and not anarchy alongside your path, to then be heard, abolishing misery among your brothers, using the technology I gave you, not to create weapons of destruction and desperation but comfort and peace.

Angels are among you, doing good deeds and being examples in society in every field. They are more noticeable in science, religion, and medicine, but they are more numerous than you think, and are there to help you make decisions in your free will, as a good teacher to students.

As I created you from my love, you must go to the school of the carnal life, preparing for your graduation one day. We can embrace each other as you came from my heart.

Messages come as the whispering of the wind to humans beings at all levels of society, but if we don't have an open mind to spirituality, that begins with the Creator, our life will be a colorless short one, as there is no future in it at all, but just to be born, suffer, and die.

19 Where Are the Inspirations?

I found out it is not necessary for anyone to inspire me or perhaps any writer. To say if wasn't for my mother, husband, wife, daughter, or whichever friends or family generations, and many times even the death of the dog is irrelevant because the inspiration comes out of the blue. It can come anytime, anywhere, and that's why I keep my pen handy.

Inspiration doesn't just flow from the mind, because from nowhere, we decided to write about this or that. Call it a gift if you wish, but for me, it is hard intellectual work, built over years. It is like learning a language; the more words you put into your mind the faster you will master it. I tried, and it works.

I have myself as an example. As a child, I was reading and writing about every subject and not only holy books. With the experiences we all have as we live, we are always taking them into our memory bank, educating it as when needed. The information will just pop up, like in today's computer, because our system is a perfect one. It is called—the mind.

I read as much as I possibly could on the sites, and it's a vast repertoire. I came to a result, but it's mine and has nothing to do to with anyone else. The mind is inside the brain, because brains without a mind are just a pile of neurons or a pile of dead cells. The mind is us; it is the spirit. If Stephen Hawking let go of his monitor and the cosmos and quanta for a moment, his "brilliant mind" could use his theory to find out what the mind is and the composition as being perhaps our conscience. Then he will officially be a "brilliant mind."

I did promise not to say a word about the Dalai Lama, considered a *Holiness*. I explain the word *holiness*. It is so noble and elevated as holistic. We used to write *him* or *he* as only a moral reference to God and then Jesus Christ, because he is respected by everyone. His parables and the Sermon on the Mount related to the absolute morality. As for his cruel death at the cross, it was proof that his mission of love went beyond the concept of religion. He learned from talking to different people, especially in New York City (NY), which is a melting pot. It is easy to sit down at the Oyster Bar in Grand Central Station, and after a few coffees and dozens of fried oysters on Rockefeller, if you are a big mouth like me, you will learn a lot from nice strangers, because this is New York, New York, as exalted by Frank Sinatra (1998).

This is my opinion, and no one will change it, especially after all my endeavors in my eight years of going through a "narrow door." I never did and never will call anyone on earth "Holiness" or "Santidade." With all my respect, I will say, "Mr. President, King, or Queen, Your Eminence Pope Frances, Your Eminence Dalai Lama," or whatever the professional people do, but *Holiness* will be for the day I meet the *Creator* and Jesus Christ in person naturally. I believe in sharing my opinion with Thomas Paine (1809).

From "What Is the Mind?" by the Dalai Lama at Cambridge, Massachusetts, in December, 2008:

"Some modern scholars describe Buddhism not as a religion, but as a science of the mind, and there seems to be some grounds for this claim."

Then, as always, a soft torrent of words on words follows; like Ravel's *Bolero*, it continues eternally without an end, until someone pulls the cord.

The average person doesn't read much, especially in third-world countries, where a person exchanges a great new book for a cheap wine bottle.

To say inspiration comes from *nowhere* isn't true, as Hawking said. The universe did not create itself from *nowhere*, as he claims in his book *The Grand Designer*, because its unsoundness proves it came from nowhere. After I read it, I felt it was teaching science for the few interested. I could find it on many sites, and there was nothing in it to keep me awake. It wasn't worth the title or my thirty-five dollars, as it had no soul in it, much less spirit. *The Grand Designer* is so boring that I wonder if all the free publicity about it was just for me to buy it. Then, I began to write this book, as I owed it to Hawking, but the general public doesn't bother to know as long as the trains just keep rolling safely on track.

As a pilot, I found out that except for the pilots and mechanics, no one else was interested in knowing why the airplanes fly. That includes the flying crew. As I questioned the general public, they would say it flies because as far as the fuel tank is full, it will go, the engines keep running, or the wings are attached to the fuselage. It's not a joke but interesting.

All this comes to a conclusion, as the average human, being in good faith, because he or she is an honest, simple, good soul, will then be mostly foolish by the unheard. They could be disguised as lambs. Being wolves, they will then have easy prey. The damage done by those evil spirits could come as a scientist spreading hopelessness or religious leaders giving fear of an unmerciful God. Once again, the list will include Hawking's universe's borders.

As with everything else, we must navigate in clear water or sky on a high-visibility day, and then the route could be safer. It goes for any of our acts related to others, especially when it comes to saying or writing something about someone's wrongs. If it is not to cause damage to that person, but as an sincere advice for the public benefit, and the intention is not to cause moral pain, then it isn't a sin.

But if someone accuses the writer, journalist, author, or preacher of a malicious intent, they do have that right, as those defaming and afflicting others also have their rights. This is earth as society is nothing more than souls on their ascent toward evolution.

A book to get to the general audience must have a diversity of facts and not just one fact. It must be like going to an *a la carte* restaurant as you make a wrong choice, but in a buffet, one has dozens to pick from, and then there will be no mistakes, but one will be able to enjoy dinner.

The average person, in general, will mostly not read the entire book or not even get halfway, but I read the book and stayed awake until I got to page 610. I slept a few hours, but it was worth it, as I was learning and entertaining at the same time. That is why I do my best to be like those "born to write" and not to fill pages with false logic. The words did not just flow onto the page but were forced onto it.

I am able to write at least two to three pages an hour, while overhearing people's earsplitting conversation at a food court in the mall (in Brazil). If I couldn't, I wouldn't have been able to be a pilot, because we must hear the communication system, have one eye on eight instruments, one eye on the map, and one eye out for birds, such as vultures, flocks of ducks, and especially Canada geese.

They all avoid our radars to send us to hell. The world knows about the miracle that happened not long ago above the Hudson River in New York City, when geese forced the plane to hit the water miraculously in one piece. The huge birds lost their feathers while being roasted.

A few years ago, a flock of small ducks were doing high-altitude training to my chagrin. I saw about twenty in a V formation, and I went into it. My plane was not a jet; they became mincemeat, and I almost had a heart attack. When I landed, my white plane was a bloody one. The propeller had to be reviewed, balanced, and polished, as did the whole plane. To my surprise, the insurance came to the rescue, and I was glorified as a lucky one. My answer, as you must know, was no one goes before their time (even "atheists" say it). I did not pay the bill, and then with the money saved, I could eat duck for millennia. I can prepare it in many ways. I am a good chef, as with anything I cook, I always say only God can do better. He never complains, because I always put him ahead.

I know people dead or alive, who are blasphemers and feel above the world. They are in wheelchairs, have no family, and have negative lives for sure. They have aged ahead of time or in front of the Creator, facing the music, because they all classified themselves as "brilliant minds" or geniuses. Thank you for a gift from the Creator called a computer. It, as everything else, came with the big bang.

When writing among people, I observe and feel inspirations as the perfect ambience, as I am not the only one in the universe. I am part of a whole family, as we all came and are coming from the same microscopic shape. Then, as I am holding the pen, my imagination and past come up, like a DVD playing on a monitor. It is called remembering. It is accompanied by all that I have learned up to the moment. It rolls into the future as our soul is reality. Reality is what humans feel, see, and hear.

Not even in the future will humans as souls be able to create an artificial mind, not even a primitive one, because the mother-plate doesn't have DNA. It is in the protein sequence or the genetic factors, produced from the mother-dark matter. All are in the big bangs, because the water we all drink and live on comes from the same fountain.

Now comes the time for the ones with some culture, not precisely as "brilliant minds" because there are no brilliant minds or people with such high IQ's, because it's all BS from the ones with inferior mindlessness, as their flesh has the same blood as pigs. They are the only irrationals compatible with the human body. If you don't know, just begin your research; open books, use the Internet, and ask the ones one step ahead of you. Whatever I write or say, I can prove. As backup today, the Internet is my angel. I said as a young man:

"If you don't know, now you know, because not to know is the privilege of ignorance, and ignoring is your problem, as the right to be ignorant, because only you can stop your ignorance. If you don't, you are still good, because time is time and will wait eternally. You are part of it, as I am also, as we all are. Some want to cross over the rainbow, and others keep their noses in the grass."

Now, in my eighties, I am getting near the end of my journey on earth. I remember the multitudes of family and friends. We belonged to the groups; now, they are mysteries, but they are kept in the memory bank of the soul. They are now spirits, confirming Lavoisier's affirmation that "nothing is lost, but transformed"—like water to steam, souls to spirit. It doesn't change.

It glorifies not one civilization but worlds or deep universes, because eternity began on what we ignorantly call the past. The past is eternal, because the important moment is the present time, and the future will always be ahead. Otherwise, Professor Hawking would be right.

No one can change or imagine going back in time or ahead of something not yet built except Hollywood for our entertainment. People make a mockery of adults behaving like kindergarteners, and why not? Famous adults have a collection of cartoons and light comedies from the fifties up to the eighties. When the stress was on, they pick, and then the film is the magical potion.

Yasser Arafat (2004) the Palestinian leader, received my book *Christ's Wisdom and the Unholy Prophet* (2002), as did the head of the Israeli government.

A few days later, he called me and spoke for ten minutes, grateful to my soul for being an angel of peace. He is handicapped in confronting such an evil force on earth, but someday, Allah would officially interfere, as he died two years later. He received the Nobel Prize for Peace in 1994.

This man was born and lived among suffering, while others sat on their bottoms in the comfort of a first-world country, not dodging bullets and eating dry bread so as not to starve. They have God as their Creator, because he like me, doesn't look at the sky on a monitor, but up beyond the clouds. That was why God gave him the Nobel Prize and to others the wheelchair.

Among people I can observe as a writer, painter, and designer. (I am not a modern art painter; I appreciate art as real as it can be, like a photo. The others are for me kindergarten distractions of the talentless ones, and not expressions of their souls.)

I can put my word and not my foot in my mouth. Look at the last pages, because anything we say and can't prove is nothing more than a theory. Theories are not realties, but as I say it again and again, children talk about once upon a time or other than what was there. I asked Arafat about his cartoon collection, and laughing, he said he wished our lives were like the cartoons.

As an artist, I can notice the difference among faces, as souls, being physical and spiritual. We are not the same in personality; as there are not two identical fingerprints, because sperm comes by millions at the sperm big bang (remember, this is a book of realities, not a humorous one, and much less a scientific one). What I mention is real. As you read this book, mockery and theories will not help you when you are confronting the narrow door of life after death or beyond life.

It isn't fascinating, because isn't like here and because there is no necessary money. Well, I cannot tell you more, because it's against the regulations, but anyhow, you would not believe it. If I describe how God is, you will want to see like Thomas (the saint).

Using all the data, experience good and bad, I have cruised on ground, sea, and air, anxious to understand the mysteries not explained by the "brilliant minds," the high IQ, and geniuses, recognized by their "fan clubs" as being geniuses. They take it easy. It is only stamped on their diplomas and not their actions. The biggest one of them all is the favorite of the media, because he is a phenomenon as the only one to have survived as a mummy, not for a few weeks, but over a half century. He is bringing a message, but not a divine one; he says that God as the creator doesn't exist. The theory of the big bang is chicken-less, not having an origin, not even from a divine chicken coop, and people believe for this reason he is being crucified by himself, because God doesn't do that to his children or anyone else either. If he did, he would not be all love.

He is sitting on a huge nail, in his divine chair, because his vision doesn't go beyond the gates of the fenced chicken coop. The chicken told the chick the universe ends at the gate. In his frustration, he sees the black holes as if they are in his backyard. He announced his dark matter is without a creator, as his nightmares began night and day.

As a sixty-five-year-old senior, I never retired, and in my eighties, I am holding up well because a pen only weighs half an ounce and the pressure on the keyboard is only 0.00001. He uses gravity in the airplane to come down, or as a senior, he loses balance, and my wheelchair is a Cessna 182 RG.

I ascended to fifteen thousand feet or about five thousand meters. On the ground, you can run it in fifteen minutes, but put on a pair of wings and try going up. You will feel like the feathers are glued to your feet. When we are up there, Manhattan could fit in the palm of your hand, and you can almost reach downtown Boston or the tip of Cape Cod visually. To the right are Martha's Vineyard and Block Island. (After all these years, I now realize it is ten times higher than the Empire State Building or over ten thousand stories high. I felt I was leaving the planet's gravitational pull. Imagine a big jet at fifty thousand feet. Now I know why they fly for a low salary.)

When I reached the ceiling for my plane (altitude by the manufacturer), I put my right hand in the empty seat and I felt his energy. Looking at the wing tip, I saw the angels, as beautiful young ladies, hair flying in the wind. It was held back, tied off. In the background was the bluish sky, because the sun had its last rays of the sunset streaking lines of rainbows, as the deep red began fading to pink. A faint yellow glow lasts only seconds before giving way to the grandiose *glowing darkness of the matter* as it covers everything beyond earth. I had full tanks and was just floating. I could be there eternally.

I felt complete freedom, being between hell and heaven. (I was advised by many publishers to avoid the word *hell*, but it was up to me. I love it because hell is for those with guilty consciences, not the guiltless. It is just fun, as it is too scary for the ones to change. It spices up any writing.)

I felt my responsibility was serious, being above ground, where my life depended 100 percent on a machine in the changing weather and hundreds of things could go wrong, but at the same time, the other side of the coin strikes back to defend our actions. Otherwise, we will not do anything at all; even so, our roof could fall in on our head.

Suddenly, a feeling of joy engulfed me as my thoughts went to the ones who never imagined how great it is to be looking down seeing how small and perfect our existence is. Our intelligence in being rational is an upgrade for our evolution; my mind went to the one who made it all possible: "Oh, God, thank you for our existence and the colorful show of your greatness, and the ability through technology we have to enjoy our world in your universe. Your glory is our glory, and it is within us. Eternity begins here, and the immensity is our home, as on earth as in heaven, as everything began on the microlevel is colossal. Thank you again and again, because we are never alone, as souls or spirits, but as a family, in relationships such as marriage, children, grandchildren, relatives, friends, neighbors, and strangers; we greet each other without frontiers, as any endeavors are part of our evolution for our glory and gives us the understanding that death is just a continuity of life. Eternity is for us as your children. Our thoughts are our spirits, because we talk to ourselves, dream, and build our lives visualizing everything, deciding and doing greatness for our own happiness and to others. It is a material, fluid matter, as to exist, it must be a composition. Everything is a composition and is created; it will never be lost, as our thoughts are us!"

I felt an energy going through my whole body, as if I had touched a naked live wire. I was in a third dimension, away from the wings, and then there was no sound, just the billions of crystalline lights overwhelming my vision. I began drifting up and away. I felt a sensation of ecstasy. How good it is to be as we are—intelligent beings!

A loud vocal sound brought me back from the marvelous world of dreams to our incredible present one as our existence on earth:

"Cessna one-eight-two RG six-three-one-four-zero, New York TCA, are you in serious trouble? Be aware you are over an intensely populated area on top of Manhattan."

"New York TCA, Cessna one-eight-two RG six-three-one-four-zero, affirmative, everything okay. I am seven thousand feet above the TCA. Why?"

"Six-three-one-four-zero, the radar registers that you are going backward like a helicopter. Can you explain it? The wind is calm, and the ceiling is as clear as it can be!"

"New York TCA, six-three-one-four-zero, thank you. Eight hundred RPM, pushing to three thousand. Request flight following straight to Caldwell, roger."

"Cessna six-three-one-four-zero, approved. Descend straight line to Caldwell, transpound two-one-five-two, Kennedy, Manhattan, Newark TCA. Contact Caldwell tower. Do it as quickly as possible. Good night!"

"New York TCA, William Moreira, two-one-five-two, maintaining three thousand RPM to touchdown Caldwell. Thanks. God bless us."

"William Moreira, very thoughtful. Good night."

"Roger, sir."

Those memories keep us alive, as we age more slowly in grace. Life is a road of happiness and also tears. It rolls with time as part of it, being eternally as the present, and it is available to all of us. Now that my physical body is slowing to age, my mind is more active than ever, because it never ages or dies. It is our essence; it doesn't belong to the material carnal body. It just uses it as a temporary but marvelous house for the spirit. Well, it took me long years of research, and I laughed and cried, but never stopping learning the goodness of our existence.

Getting home about 4:00 a.m., I felt like a teenager. I added to my quotes, my experiences with the remarkable night and what I had written while with my wings standing still, between gravity and neutrality, in a spiritual zone. My *Parker 51* doesn't respect heaven, hell, or anywhere, especially shopping centers in the food court, as in Brazil, it feels like it is in a chicken coop where the feathered ones are fighting for the corn.

Now, at my turning point, in my eighties, it puts weight on my shoulders. I changed my fifteen thousand above the clouds, as I had my eyes on God, angels, dozens of instruments, planes, and clouds. I was worrying about hitting the first turbulence and worse yet, being face-to-beak with birds, especially ducks in high and low altitude.

Now, at sea level, I was cruising. My large beachfront windows offer a view of the thousands of people. An hour passes by. I have my eyes on God while listening to a great no-ads classical musical station. My spirit never ages or dies. I just keep doing what I have to do.

While on the ocean cruising, I felt the light movement of the waves. Being pampered by a new universal family, I thought my worries now were to keep one eye on the ocean and papers and the other one on the hundreds of passing passengers. I smiled and said hello while words exploded onto the paper with good intentions. God is always in my heart as time goes by.

A few days later, after my last landing, I was considered by my family and friends to be crazy. Being a private pilot is to fall from the sky or explode into pieces on the concrete, with or without faith. Limits are limits, and I had pushed my luck.

After breakfast, I entered the FAA office at Teterboro Airport in New Jersey, announcing I was giving up my right to fly. I asked what I should do. They all stared at me as if I were an alien. They said all I had to do was to turn around, go home, and put the wings in the closet for a while, as everyone smiled.

When they noticed I was serious, they explained I had to sign a statement giving up *my privileges* to fly as a pilot. It was as simple as that, but I was the first one to do this act. I deserve the Nobel Prize, as the ones with teething troubles will fight to the end not to give up what they love and that is flying.

As I explained to them, I had made a promise to my deceased daughter Carol and also the whole family that after a thousand landings I would put down my wings and quit this spectacular hobby. I had done forty-two extra night landings, as my wife and grandson counted from my pilot's log. But my problems weren't just flying the airplane, because I became its master, a daredevil. I was not respecting it as a delicate bird. I dropped like a shot duck and then landed like a swan. No runway or grassland was too short to land or takeoff, and when this attitude shows up, you are the first one on the list of the to-be dead, aging was also taking its toll for such a delicate reverie.

At the FAA office, they applauded my short speech, but once again, they wanted me to think about whether "to sign or not to sign," because after I put my name on the paper, there was no going back. I could fly anytime, but with an instructor (like a student), and to get a new certificate, I had to go through a regular test.

Once again, I picked up my *Parker 51*, and this time I said thank you to the Creator for my short-lived career as a pilot. I had gotten spiritually closer to him, opening more of my frontiers of learning. Visually, I could feel my faith in him, and I didn't become a pile of debris or a funeral torch pile. I satisfied my childhood daydreams of soaring like a bird.

I left my wings and feathers at the FAA office, thankful for this great organization that so gracefully did their responsible job with God as their guide. I felt like crying as I held in my emotions. A few days later, I received a memorandum from them:

> WILLIAM MOREIRA (Canno) journalist, writer, author, came to our Office at TETERBORO AIRPORT Branch, New Jersey, on his own will, without any pressure, in front of eyewitnesses and physiologically apparently well, he surrendered his PRIVATE PILOT CERTIFICATE and was advised if he wanted to solo again, he must do all tests for getting a NEW CERTIFICATE.
>
> William Moreira (Canno) affirmed in writing to know it, as was his intention to surrender for sure his PILOT CERTIFICATE due to his senior age, and he satisfied his wishes of conquering the sky as a childhood dream, as he did well reaching the ten-point mark, as a proud member of the PRIVATE PILOTS ASSOCIATION OF NORTH AMERICA.
>
> Richard Smith
>
> FAA Inspector, North East Coast Area of the United States of America.

20 The Material and Intellectual Evolution of the Human Being as Represented by Stephen Hawking and Interpreted by the Author

Evolution is perhaps the most used word in science, and it has many meanings. It is nothing more than an upgrade for better, rational human beings, irrational animals, as vegetation, including the cosmos. It is just scratching the surface.

Science uses evolution for any item, including man-made material, such as machines and medicine, and the microscopic world, including cells and germs, as the infinite creation.

Nothing is really modest in our existence, because if it wasn't for a single particle, there wouldn't be any existence, because it is the accumulation of single grains of sand that make a beach. The dark matter is what the universe(s) came from and the material it is composed of is like the grains of sand at the beach. Now, take a deep breath, because we must accept that, not as a religious choice, but as the only way, and be happy. We are all in the same wagon.

Our technological evolution never upgraded so quickly as it did in the last one hundred years. I was born in 1933, and in eighty years, I saw the leaps in its evolution as a blessing, if I can use this expression, and why not? I do say it is a blessing because as this evolution continues, it is an unknowable power. It is done by nature, and we have no control, as we have no control of our birth and dying process. We are just a part of it, and beyond science and religion, you and I are mortals. We hope our Creator will continue our intelligence in the spiritual form. We knew it since we consumed raw meat, because we did not know what do with fire.

As it became dark, before sleeping with the chickens, we stared at the sky and extended our arms. Monkeys jumped, trying to help us, as we tried to reach the stars, knowing his eyes were watching us. Since we began talking and put writing on paper or even other material, such as clay or stones, we have been seeking him as a true consolation, as in the past, present, and the eternal future. He is the only creator and will not show up to the creators using their free will to say that the marvelousness of his creation just popped up from nowhere. The ultimate question is the beginning. It is just an eternal mystery, and the brilliant mind or genius is on the same level as the ones who glorify him.

Technology is a double-edged sword, because it can be used for military killing and destruction or by disturbed minds, as they do not fear a Creator, because there is no Creator. Consequently, destroying human lives is like killing chickens. They can even destroy earth. As the sequence continues, now there is Kim Jong-un of North Korea. As one idiot leaves, another one comes into power. Hitler did not leave any successors.

I will admit someone has a brilliant mind if the genius dedicated his time while here as a leader in a third-world country and earned the trust of the people, making his heartland a paradise on earth. Chavez (2013) tried it in Venezuela, as the father of the poor, like a Robin Hood, taking from the rich and giving to the poor. It did not work, because education must come first, or the stealing will continue.

Great religious and political leaders, such as presidents, dictators, kings and queens, prime ministers, and even movie stars, singers, scientists, industrial entrepreneurs like Henry Ford, and so on arise. Today, Stephen William Hawking, using his theories, programmed himself as an image of being a brilliant mind, a genius with a 250 IQ. The common people as children follow any fairy tale to its destruction, creating a universe without a Creator, as an incentive for the devil's minds to have no respect for someone to answer to, because after death there is no life. There aren't more universes.

Hope is having our existence as a "chip" off the Creator, the one and only. His beginning is a mystery as it was a mastered creation by a brilliant religious mind. It is called *hope* and stops tears, giving the incentive that life is worth living, despite all the suffering. The colorful rainbow comes after the hurricane, and there is writing on it: "Our beloved ones are now with him, our almighty God, and we must continue living life, remembering good, because the sun is as bright as hope, and the seeds are growing."

In the last century, better yet, in the eighty years of my era, the jump in technological advances went so high it reached the moon, but luckily, it missed Hawking's black holes; otherwise, there wouldn't have been any satellites.

I saw the birth of the jet age and computers, TV, movies in color, perfect photos, and, as a pilot, the GPS come to be a reality. I was in my plane on a foggy day only few feet from the ground. The training was to look at the instruments' guidance as God doing his work.

The United States is the number-one country and will continue being the first country on earth; the people work as one family and help other countries as in the Tripartite Pact in 1941. Adolf Hitler, Benito Mussolini, and Emperor Hirohito, as puppets for a *Satanic militarism*, cost America dearly in blood and industry. They helped the oppressed countries save our planet from the pact of evil.

But, as always, the United States of America is the one putting a stop to evil. The great nation is again putting its intelligent power to stop the militarism evil of the puppet North Korean Kim Jong-un. Even his picture is an eyesore, as are all the agents of the dark spiritual world. Hawking should look in a mirror for a surprise; even as a teenager, he was a soulless man.

This blessing will be like this eternally, as long they keep on the monetary system: *In God We Trust*. Americans are the number-one churchgoers on earth, as hope is part of their hearts. They received the number-one anti-God promoter, Stephen William Hawking, as he lived for years in California, having twenty-four-hour care for his disability. Everyone knows he even tries to kill philosophy. Hope is still on a rock pedestal, because our planet has more good souls than bad ones.

I came to New York City on January 18, 1956, as a young man of twenty-two years. I was a journalist there for a thirty-day visit. I felt the cold winter and snow, learned English, and enjoyed the good beef and fresh Maine lobster, with my good camera (Minolta) and my *Parker 51* pen. I felt a country, and I knew its history and its people by heart. It was not because of monetary difficulties that I kept staying. It was my admiration for the country's integrity. I had and have the honor to be part of it.

In 2002, I officially sent to iUniverse.com my first book *Dr. Fritz: The Phenomenon of the Millennium*. It had more than 440 pages and 180 photos taken personally by me. It was heavier than a suitcase. A few years later, it was a DVD, and now, as a PDF, it goes straight from computer to computer, officially free and safe. It came back for corrections and approval, as a grace of God.

As always, Americans are pioneers, especially in commercial aviation; Boeing and Douglas are always ahead. There were others who jumped on the wagon, but now, it's called progress. Boeing won the competition and was granted funds (Ronald Reagan administration) for research. They built a supersonic plane with at least two hundred seats for long distances like New York to Beijing in about two hours. It is top secret. I read Boeing invited Douglas and Lockheed in partnership. There were technological difficulties, especially related to engines. I hope it will be a reality soon. I can materially witness another human leap in science, because I could do it as a ghost.

Our worries must be here for us for a better life, not dedicating a lifetime looking to distances. Only the most powerful telescope can reflect in our retina, but our problems are right here. We can make a difference in our material life. Medicine still has far to go in giving us more help, as do agriculture, energy, and climate control. Hurricanes not only jump over our cities but destroy them. Roads are still unsafe. Planes can fall and do. Ships sink. Cocaine is destroying our populations. Meanwhile, a Godless, soulless scientist is receiving presidential honors, not for saving the planet or even himself from his heavy cross. It is like we are receding in time.

I am happy to have been born in our world in God's universe, at the right moment to see human beings getting more intelligent, as some go above the average mark. They bring more evolution in our progress in generally giving our lives comfort, as it is programmed, and not to profanity taking hope away. Life has no future in opening the doors for the anti-God. Worse yet, one proudly promotes a Godless society as he satanically manipulates the media. This book is more than a book; it is a warning from the Creator. Today's youth now adore Hawking, as he reaches every classroom and media outlet. In the third world, he is known as the man who defies God in the open.

We are the ones in need of material and spiritual evolution. They are like bread and butter, because we are created and God is the creator. Would we be anything at all? The evidence of God's unchangeable laws is there. "You do wrong, and pain will follow." Hawking's being frozen on the outside and alive on the inside. He doesn't recall it is a lesson from God. He says his condition is not God's punishment. Neither is it a curse but nature doing her job. For a half century, he has refused to admit there is a Creator, while sitting on his "nail," alive to feel it.

Men believe they are being good and advancing in evolution, materially and spiritually, because no one is shooting arrows at each other. They are right, but they are still killing each other at a distance with bullets. Cities now are not conquered, but exploded into pieces or being incinerated by atomic blasts. Better yet, the arsenal for biological warfare continues alive and well. When the sky doesn't drop water or drops more than they can handle, they all say God is upset. Is he?

I just got this one down somewhere in the memory bank, as President Reagan (2002) said: "Freedom prospers when religion is vibrant and the rule of law under God is acknowledged."

Interestingly, sometimes, people say, "Why should I read a book from someone who does not have a famous name?" When a book is "discovered," it means someone in the media sees it as interesting, and it could sell more than one from an already well-known author.

Also, not every dish the chef puts on the plate is a masterpiece of culinary art. Hawking's other books did please many people, but his last one *The Grand Designer* came up mediocre. It was sold because it was in the shadow of the previous one. It also damaged any other future ones. It's like the movies from the Jaws series. The first was a success of emotion, but the second one was a dud. My books with spiritual messages based in reality are becoming dust, as millions of others, with nonsense writings go on the *New York Times* best-seller list.

The main reason is that we are not in heaven, but in a school for souls to learn love and become more intelligent. Who's doing it? Don't worry; sooner or later, everyone will have the answer—as soon as we all leave or abandon our carnal bodies. (I don't like the word *death*. It sounds too dramatic, as it comes with our birth certificates as a bonus, saying we are here on vacation.)

I bought books and DVDs explaining how it works, as well as computers and chips, but for me, nothing seems logical, because inside the chips there is nothing more than a few grams of solid metal. (It is porous, like looking through a powerful microscope.) But so what? How does it contain millions of bits of information in it, and how do changes reflect on the monitor? I take it as a gift from our Creator. The same is true of our bodies; we began as microscopic bits called zygotes, which trigger a soul. It becomes us, and no one knows how it works or was created. It works or you wouldn't be reading my book, nor would I be writing it.

That's the reason I believe there is a brilliant mind; I call him the Creator or God. It is more than a reason for the *brilliant minds* and geniuses to bow their heads in thankfulness for existing as human beings, feeling and enjoying the marvelous creation and not as an irrational chicken that is looked upon as a meal.

I knew Robert Barroca in Rio for the last thirty-five years, and now, at the age of ninety-two, he's still a genius in electronics. He began by fixing broken-down radios. Earlier, when he was in his teens, he became an expert following the evolution of hardware and did well financially. He owns a great apartment in South Miami. He came a thousand times to buy parts for his business and take upgrading courses. He is so good in this difficult line of work, because there is always a need for those gifts of repairing gadgets. Even now, he is sought by militaries to help with the airports' problems with radar communications. He resolves, sometimes in minutes, what seemed like it had no solution. The technicians ask him, and he finds the problem right away. His answer came with the help of angels.

He says he also was born at the right time as he grew up with modern technology. He always says things work, but no one knows how, because as a good Catholic, he affirms it is God's gift, like life itself. Many times a day, he mentions God: "If wasn't for God ..." or "Thanks for God" or "God willing."

I met many such faithful people; they all are happy with life and healthy, like Barroca, who is in his nineties. They bring hope to everyone's life, and as a reward, none of them are sitting in a wheelchair. A few days ago, I dropped in at his home in Ipanema Beach for a light lunch. I gave him one hundred pages of this book to browse. He gave his opinion. Smiling, he said, "Hawking should have picked a profession to help us right here and not wasted his valuable time seeking black holes, including the ones he is in, but in his mental illness, he killed God, trying to be a

God. He gives the logic of creation, while sitting on his nail as you wisely create the parable of *The Nail and God's Dog*." Then, he added that it was a good book, and even if it didn't sell, I at least did my job. He believes I am sincere and God used me to alert the younger, newer generation. We must admire the Creator, as he keeps taking notes of our behavior. In the beginning, he gave to each one of us the deservedness according to the unchangeable law of cause and effect, and then he invited me to a cute little centennial Catholic Church for the Ritual of Blessing the next Sunday, and I accepted from my heart.

My dear friends, sometimes things that occur in our lives seem like foolishness, but they aren't. It is easy and difficult, like "Dante's mask." One side is smiling, and the other is crying. However, everything was orchestrated by a Supreme Intelligence; there is no other explanation. Everything functions as it is meant to, beginning with the spectacular origin of our carnal bodies. I say *carnal*, because into it goes our spirit as we command from the brains enclosed in our skulls. Our souls need to be liberated as spiritual beings; they must come out of the created casing and bust out as a divine, colorful butterfly.

Dante (1321) wrote in the *Divine Comedy*: Concept of Contrapasso:

"The sin equals the punishment; sinning is its own hell as it destroys the very life of the sinner."

As I promised Stephen William Hawking a half century ago, my first printed book was dedicated to his good deeds. I will not put it on my rock pedestal but mail it to him, hoping for his own soul to benefit, as well as those of everybody becoming Godless on his Satanic theory. It would fill his monitor, as his secretary reads it to him, beginning on the first page. It is an open page, as we read and learn from it. After that, he can die in peace, because he will find his soul. He could have as a reward more years of life, at least be able to move his body while savoring fine food and smiling at the touch of his children.

Now, back to Barroca's home, as we walk to his and my favorite spot for coffee, his gorgeous kitchen with the view directly on the Atlantic Ocean; its pounding waves hit the curvaceous Ipanema Beach. My energy comes from him, and not a casual wind; because in a no-wind day, the waves are the same, night and day. (My idea is that it is because the rotation of earth is as smooth as it can be; it has a vibration causing the oscillation of the massive body of water that is the ocean. As it affects it, the cruise ship continually crosses it from America to Europe; it is a notable phenomenon.) When the wind hits it, the small, continuous waves become huge. They came from the small ones, waiting to be pushed into a demonstration. God is around to keep the boats floating, as the helicopters are the angel's wings. They battle against natural force, because they are on a rescue mission of love.

I was getting up to leave, and as always, he had a light joke to keep us smiling. He said I am a genius with the pots and pans, but his was a brilliant mind. He holds the knife and fork as a gourmet. Life is beautiful as it is. Why then do we fight only to lose against its current.

I took a cruise in the fabulous *Ms Allure of the Seas* (Royal Caribbean) leaving from Fort Lauderdale. It was 220 tons and 5,400 plus 2,384 souls. It makes human beings feel like they are God, but as soon as the majestic colossus moves, we all know it because we begin feeling we are not on safe ground. Gravity tries continually to do its job, to pull everything into the center of the planet, and any floating might is nothing more than a rice shell, reminding us we have a Creator. Any light balancing the merry passengers just whispers, "Oh, God." It's an assurance that everything will be well. Now, to take this belief from the human souls must be the work of a *Satanic* mind and not a brilliant one, much less a genius.

I love cruising—the comfort and food and entertainment. We are surrounded by uncertainty. It is part of being here at the beginning of our existence. The sailing is going smoothly, and then come devilish contradictions to change things, as it took *Costa Concordia*'s alcoholic captain and not an iceberg to sink it.

Our evolution as human beings is a slow but sure process, as materialism comes many times ahead of morality. The ones using their tongues negatively are doing more damage than a nuclear holocaust. Those with a ferine mind and tongue are the first to succumb to their own actions, as I prove in this book. He's the star. I gave him a half century to find God and leave us our passport to eternity called hope, but I didn't waste that time. It is because God is never in a hurry as time is as eternal as he is and we are as his offspring.

As the old saying goes, the more we live, the more we learn. Those negative ones who love to spread terror, their reprehension comes in all shapes and colors, as bad spirits are being called Satan or whatever. Now, they are revealed by the Vatican as spirits off their orbits, attracted by sinners and keeping the Vatican's exorcism department very busy and canceling any days off and also vacations.

Deservedness is in the evolution of our world, and it will be the end of many mysteries for science and religion in their research, including gravity and how to use it for transportation—levitation. Ghosts have this privilege. I would like to have the option of hovering over the filthy potholed sidewalks all over Brazil, as another off-orbit country and include its contaminated, grimy bathrooms as well as its sandy beaches. It is in my nightmares as a clean soul.

Log on to the Internet, on to sites like Wikipedia or others; there is great information available just by clicking the mouse—cultural teachers in front of your nose anytime. Information opens the doors of intelligence as evolution; the more you know, the more you can distinguish the difference between a dullard and a genius. Evolution is a slow process, and the reason is because eternally we belong to time as it rolls on; it is a part of being born. If everyone with a computer would use it for only one hour, click to Wikipedia seeking to know reality or to *National Geographic* and the Discovery Channel, we would be on a better level.

I never went anywhere at a home, even mine, when they had those channels on; rather, it was a drama, ball game, or some mediocre show or comedies where they do the laughs for you.

Evolution comes with the help of mankind's dedication to science, medical research, religion, and the other professions derived from it. No job is better than another, because without the ones cleaning up the sewer systems, we would officially be microscopic creatures' meals; without the ones in the field planting vegetables, there would be no food on our tables. Without the bus drivers, pilots, or ship's captains, there would be no international merchandise and grains exchange. We need the cooks, painters, janitors, gravediggers, police officers, and so on. I say the few brilliant minds are so many that they could illuminate our planet. Why call a particular one a jackass, while millions stay in the dark shade?

Now, for generations, everyone has heard about the American Thomas Edison (1931). He came to illuminate our paths with the lightbulb, and more than one thousand inventions we needed so badly. He didn't brag about his IQ. No one called him a brilliant mind or a genius. As for Christ, he conquered the world with humbleness, not as a fool, because he guaranteed material wealth and vanity don't follow the soul to an afterlife, but good deeds do.

Henry Ford (1863–1947) made a giant leap in the automobile industry. The assembly lines brought comfort for everyone, as prices made it affordable for every pocket, while paying a decent salary for the workers, including fair housing and schooling for them. Real brilliant minds give hope on earth and not illusions beyond the galaxies.

Going beyond the horizon as a number-one entrepreneur, he fought as a pacifist to stop the First Word War conflict in Europe (1914–1918). It brought death to more than five million soldiers and civilians alike, darkening the horizon of pain, as prelude for the big one. Hitler came onto the stage of death and destruction.

This calamity provoked by godless men was just sixty-eight years ago. When it ended in 1945, I was twelve years old, and Hawking was three years old. He was a toddler with a pacifier to calm him down. His mother, like all mothers, dreamed for him to save the planet, but he chose the wrong path. He began paying the price.

Today, even those who were teenagers at the end of the terrible conflict have never heard of it, as it is with any event with all that negativism. A half century from now, when someone asks about Hawking, they will say, "Stephen William Hawking? Does he play on the Dallas team, or is he a pilot in the Star Trek fleet? Or, is he the one who invented the robot housekeeper, Hobo?"

In 1949, I jumped in a Ford model 1938, as a seventeen-year-old kid. My father had bought it, and while he was traveling by train, I just drove it. It was simple—press the clutch down, move the stick forward, release the clutch slowly as you press on the gas, and bingo. In Brazil, at that time, if you had a car, the street belonged to you, because only the United States had the technology to produce and supply the new need of the public.

I love the sound of the engine of propeller planes because it reminds me of the elegance of the past. It has life when the propeller swirls. Also notorious were the steam trains. People used to line up to see them passing. I was a child seeing the engine puffing a lot of black smoke, spilling steam from everywhere. The giant wheels connected to each other and made a sound, saying it was all might and to keep the track free. I was petrified, as it was alive.

I became a collector of miniature trains moving on Lionel tracks. I still have the powerful *Big Boy* in my living room as a reminder of when I looked at life as a young man. Life is beautiful yesterday, today, or tomorrow, because we are the ones who do it in our free will.

Louis Pasteur (1895) was a chemist and biologist, who invented the process of pasteurization. A father of microbiology was more than a brilliant mind or a genius but a hero to humankind.

Who hasn't heard of penicillin? Alexander Fleming is credited with discovering penicillin in 1928. It saved and is still saving millions of us, dominating the microscopic enemy, which eats us alive, anywhere, anytime. We feel but don't see it. He came to us, not as a brilliant mind but as an angel, because millions of soldiers were dying in the Second World War, more from infections than bullets. They were dropping like flies.

We need brilliant minds in medicine, chemistry, sociology, and so on, people who feel our planet materially and spiritually. We called them mankind's heroes, defending us with the help of a microscope, while using their time off to play with the telescope and see the show going on billions of years away or enjoying the imaginary world of spaceships, as I did in my early years with planes and trains.

At his funeral, the world came to pay their respects with applause and tears of thankfulness. He brought hope and not imaginary theories and a negative godless existence. Hawking wonders why he hasn't been selected for a Nobel Prize. Space theories on cosmology don't save lives, put food on the table, or help earth avoid the destruction of the present real enemy called infection and the evil alive in the form of atomic bombs, a fanatic's toys, as now North Korea and maybe Iran are pulling the strings. We are ants and not human beings.

In his first speech, the new pope (2013) said loud and clear that hope of a merciful God is all we need, because without it, mankind would be in darkness. I hope for his benefit that Hawking heard it. Now he is fighting against the current, as it only brings losers.

According to the past, as in his biography, Hawking wasn't a brilliant mind as a student and loved to drink, and then when he fell down some stairs, he lost his mind temporarily. That was why they did a test to see if his IQ had been damaged. Spiritually, everything became wrong as an effect of the cause. He woke up from the fall on the dark side of the memory bank, and it kept him in a full campaign against life, as he unmercifully took our hope of a continuous life after death. He does it openly backed by a media, where the godless are his promoters.

The majority listen to the pope with or without religion. At his inaugural sermon, more than 132 high-ranked officials of countries were there, including Dilma from Brazil and Christina from Argentina. Coming back home, Dilma said the pope offers hope for all of us to live in peace, while helping each other for a better fraternal life, as it is God's unchangeable law.

The evolution of the mind is also the soul, because it is nestled in the brain, commanding the carnal body through the nervous system. It is like how we manipulate robots with remote controls, like string puppets in a show. That's why when the brain stops any action, we are dead, but as you now know, death doesn't exist. As the electrical energy of our carnal body leaves it, our fantastic human journey ends and the body becomes useless. We as ghosts with all the senses will float away in the direction of other worlds. It is the infinity of our imagination.

You can take it as a fantasy, dream, or faith of a continuous life or just another ghost story by Hollywood, but either way, as Hawking's "brilliant mind" believes, it is not a spirit in command of all the existence; it is his inert physical material carnal body. He just jumped on the wagon of unhappiness. I, William Moreira (Canno), embraced hope as my light in the dark tunnel of our short journey on earth. I enjoy our huge global family, holding with love and not blaspheming our Creator, as we are intelligent beings in our free will. We have our world in his universe. It beats staring at mud in a seven-foot-deep hole in the cemetery, waiting for our cadaver, which will be there at any moment as the final bit of energy called the mind vanishes. Its intelligent mind, for Hawking and his fan club, is just part of the trillion neurons. (Hawking did the counting, as he has all the time, because he cannot even go to the bathroom, but to be sure, his anti-God propaganda is doing well.)

21 Who Am I to Talk about the Reality of Hawking's Negativism?

His *Brief History of Time* sold millions and was translated into more than forty languages. Hawking compares it almost with the Bible. His supporters hadn't expected it to go beyond the limits. He claimed it went up to their expectations. People question him, "Where did we come from? Why is the universe the way it is?"

He affirmed it and gave logical explanations from a scientist, but I read it four times in a few days, seeking what the advertising claimed was the work of the number-one genius, the father of logic when it comes to God and cosmology, and found only the BS of a mind atrophied in a mummified body. He became rebellious about being created from a zygote. We all know, as common sense, that it is more than that. It's the Creator's way and works as everything else under his supervision does.

Stephen Hawking is getting away because Thomas Paine isn't around to answer him as he did all the atheists. The wrong political leaders, kings and queens, dictators, and party members invited him to the French Revolution to give some guidance. He said: "Kill the kind, but don't kill the man."

But the minds beyond the guillotine's trigger didn't have souls; the blood gushed like Niagara Falls and included that of the king's wife, Marie Antoinette; the two children; and the head of an important man, Lavoisier, the father of chemistry. Hawking describes the species *Homo sapiens* (human beings) as being monkey-like.

"French don't need chemistry but soldiers."

As I say, "Human kind does not need brilliant or genius cosmologists to tell us when the sun is going to cremate the earth or a black hole is going to gulp everyone forever, but Hollywood payrolls employees to imagine our spaceship voyager movies, as Americans are the best in the industry."

Now back to Thomas Paine; he was an English citizen from a poor family, but it doesn't matter because you are not your family, as you are the only one to use your IQ to open your door. My family was poor, but at the age of eighteen, I had the grades to get into any university, but I chose the 180-degree opposite on the road of Windom. At nineteen, I had my journalist credentials, and as the journey then became too boring, at sixty-four, I earned my pilot's certificate at the busiest airport for up to medium-size aircraft in America, Teterboro, New Jersey, under New York, La Guardia, and Newark TCA. Bingo! Another winged daredevil as Icarus was born, but I exchanged the wax for aluminum, as I could then see God closely.

I once did eighteen thousand feet on a cold night. As the air became icy, I swirled the propeller to a standstill of eight hundred rotations per minute (RPM), not losing altitude while I chatted with our Almighty. (Before you ask, the minimum RPM with full flaps is about 2,500 before you begin losing to gravity.) As I talk so much about aviation, here's the answer to the question about how much gas is necessary to take off and land:

"A full tank to go up, and as to come down, an empty one, because gravity is free—and it doesn't pollute either."

Here is a quote by Hawking as he became famous for his *Atheist Club*, the ones gone awry while seeking any understanding of "why life is like this" followed by my comments on it:

"We are just an advanced breed of monkeys on a minor planet of a very average star. But we can understand the Universe. That makes us something very special."*

The quote is his. No one in history ever said such a joke as offensive, because Hawking is not Bob Hope. He demonstrates his bottom is really sore and he can't scratch it. As miserable as his life is, he wishes we all should jump in his wagon, but I am not blaspheming our Creator, and as my reward, I can do all my scratching, jumping, flying, eating, dancing,

* "Stephen Hawking," Wikipedia, accessed December 10, 2012, http://en.wikiquote.org/wiki/Stephen_Hawking.

writing, traveling, and so on using my free will and am happy to feel and know there is a splendorous afterlife for all of us.

I will not comment on any more of his quotes, because I won't waste my time in mediocrity. There are billions of things to learn or to teach worth our time, especially going to a foreign country, because on our *minor* planet, we would need to live two million years to sightsee just the capitals of all the countries while savoring the best they offer—not sitting down on a punishing "nail" as a heartless anti-God.

As perfect as our existence is, God could never make an irrational being with only the instinct for survival, a body fitted to jump on trees, have sex in the open, and eat bananas, as its DNA was programmed for, like all the other species to then be a *human being*. Our body was programmed and classified as *to be intelligent* and have the capacity to seek him, as seeing him in his marvelous creation. We can create on the created, like this book, and have the feeling of love and appreciate beauty, not just stare at it like a hobo.

Yes, when he decided it was time for intelligent beings similar to him, as I adopt from the Bible; he sculptured as a semblance of himself a human being from a mount of clay. He energized the being with his energy, and bingo, here we are, soul and spirit giving us the free will to go up from eating raw meat to a delicious oven-roasted prime rib of beef and everything that goes with it.

Hawking feels like a monkey, because monkeys are like any other irrational beings and will be eternally, but he will be free from his monkey mind and body, the day of his last breath. Taking his nose from the monitor, he will face God, and he better hold his spiritual breath or he could reincarnate in India as a rat and not in his family tree *Catrol vancliechin*, as he belongs to the top branch *Macaca fascicularis*, as macaque, the only monkey in the monkey family capable of seeing black holes. Hawking called us all a breed of monkeys. I found out about his excitement about being godless, as he was the first monkey where his father was a gorilla. The laboratory picked the wrong sperm and his mother had an artificial insemination. Please, have a nice day.

22 December 24, 2013, Arrival at Rio's Port with a Book

As I passed by the midnight buffet at the main pool, I filled up my coffee mug and then went up to the top deck, where the wind began to pick up. I decided not to write but to mentally put my conscience in order so as to confront the abnormalities that for sure will surface on commentaries from those who just condemn, doing nothing for the benefit of humanity.

Worse yet are the ones doing a lot but doing things only to their own advantage. The hell with the other—they are saying, "All for me, and nothing for you, as are the salaries in the third-world countries."

At my age, we are more than ever thankful to our Creator for having put so many miles and years in our bodies, as he must have a reason for it. Mine is to expose the unkindness of the lost ones that comes from the top and make shreds of the ones incapable of defending themselves against being called "a breed of monkeys" by the one who gave us the *universe egg*. For gratitude, he spat in his face as Judas did to Christ, but he did as common sense and nailed him to a chair, not to save souls as Jesus did, but his bottom; as he brags it was the germs, but who created the germs?

I am up in the veranda of the cabin, on the seventh deck, seeing the sun just facing us from behind Rio's landscape. The port is less than a half hour away in the historic downtown Centro.

Holding the over three hundred pages from my *Parker 51,* just like a mother holds her baby, as once again, I know the writing gets me closer to God, I ask him many questions, as does everyone, but I sense his answers. I keep writing as the communication channel among us.

Cruising for ten days, I felt like I was in a paradise in the company of three thousand souls, happily as a perfect family. Food matched the expectations, the weather gave us a break. I also gave the captain my last book, written while I was on *Concordia*, and he sent me a top-rate bottle of red dry Spanish wine. I will keep it until the day the book is published.

As I mentioned so many times, this book is not to defame Hawking, the Catholic Church, or anyone else, as the facts I present are well known even by the Brazilian Indian tribes lost in the green inferno of the Amazon, the aborigines from Papua New Guinea, and the ones lost on the freezing tundra of the Russian no man's land.

Reality is not two-faced, but what is in the light, many hide from it, making themselves believe in what is not real, as Hawking affirmed us as being from a breed of intellectual freak monkeys. I mentioned the result of my research on his family tree; I found him on the top branch, not as a monkey, as being inferior for a brilliant mind but at the top as it belongs to the maximally great ones, the macaques.

Hollywood used them as recommended by Hawking for the successful movie series: *The Planet of the Apes*, as his relatives like him can walk, talk, and do as humans do, making it easier for the producer, except naturally to have a God, but a gorilla is on the pedestal of their temple.

As I went to the breakfast deck, I met a few hundred of the ones leaving the cruise ship and I told them that now they are my family, as the book is for everyone and not a particular group or religion, reaching hearts for a better life on earth, where equity brings solutions based in our Creator. The godless ones should keep their months to themselves, as they must pay attention and see that most of all the calamity is from the ones without fear of having to answer for their wrongdoings to the one and only who originated our existence.

I wrote this book for the first time in Portuguese, because an editor wanted to publish it, but I changed my mind, translating to English and letting iUniverse.com do it first, as I feel it will be properly done. I feel in my soul this book needs to reach a wide public, and the English language is the universal communication system. The public is avid to know, especially in the United States, a number-one country, because wisdom brings evolution in the form of understanding in our carnal world of challenges.

My next step is getting a cab to go to Copacabana Beach, as it is only fifteen minutes away. There, my computer will take the task from my *Parker 51*, and then we go from there.

23 February 24 to March 3, 2013, on *Costa Favolosa* to Finalize the *Big Nest*

By now, all of you must know that there is no effect without a cause, as affirmed Stephen Hawking, the number-one brilliant mind and genius. God needs him so much on earth he is keeping Hawking here. He is passing a half century chained to a wheelchair. He doesn't want him there yet, as he learns some lessons of humility.

On the ill-fated *Concordia*, I wrote *GOD! The Realities of the Creator.* It was in its womb on a thirteenth day in 2009. I was in the same spot as I am now on the majestic *Favolosa*, which took the place of the great moribund cruise ship that vessel was, with comfort, food, and entertainment.

Interestingly, the *Concordia* sank on a thirteenth of 2012, as I began writing it on the thirteenth deck, aboard it also on the thirteenth of December. The book commemorates its thirteenth month of publication. The *Titanic* sank one century ago, but what is the highlight? No one cares; it's almost like the irrationals. The average human worries more about what is for dinner, if the payroll will rise next month, and whether the little pig in North Korea is going to press the green button to roast South Korea, and then becomes a *big pig roast barbecue* by the Americans, as they are the best at it!

As I entered the vessel, I felt guilty comparing *Favolosa*, as a great cruise ship, to *Concordia*. It was the royal one. As I closed my eyes, I felt like I was back on it. It was as if it were yesterday, and then it was just history, as it is one of the unchangeable laws. No one will ever warp time, because time is also included in it, just like gravity; it's the present, always moving, like the earth's rotation. It doesn't matter. It is always moving; everything is timed or would be in perfection. Everyone came from the union of a father and a mother, just like the eggs without being fertilized would not become chicks, now and eternally, the right way, as it was planned. Every woman wants to have children, because it is an instinct. It isn't necessarily planning, as the free will stops on the barrier called the Creator.

The next hour is always a surprise. I have hundreds of them, good and bad, as part of being here. I thank the Creator my mother did not believe in abortion, or I could have been in the long line trying to beat other billions of fellow sperm to win the race to the ovum.

Some souls are born on silk, while others, like Jesus, in the ghetto, but I see it as a challenge to all of us, as an IQ test. We go from poor to rich, and that is when the common sense goes to work. As you make your life, it seems like it is a throw of the dice, like Hawking says. God throws his dice. I now know why; it is because he is God, and being so, he won't allow Hawking to find out where the dice are and then play god. The way he talks, he feels he is impressing the public, but piety is all he gets. Did he regulate his manners, as the way he softly talks like being a victim with his right theories?

Right or left, what counts are the performances and the slice of it. There are two groups, the ones who think they are right, and others who confirm. They are the good ones, while the others are the rebellious ones!

Talking to a senior for few hours, I felt his culture, as also he chose to speak in English. He spoke well, not as a show-off, because when you know, others notice the naturalness of it and how much more comfortably it flows, than does that of the ones forcing what is not there. It makes the difference.

He read a few pages of the *Big Nest* and then gave his opinion; he said it is a mix of spirituality with a touch of science, more philosophical in a way. It offers a touch of help into the incertitude of human lives, those who exist in unbalanced doubts as to what is going on. Everyone is uncertain of an existence, and they all seek solutions and answers. There are no solutions, and they all are in a sinking ship without a way out, because death is real and the afterlife is a vessel at a port. The vessel is still imaginary. We all emotionally dream of it, or are we already in the dreamland, as spirits?

This simpatico or charismatic senior admitted that I was saying what thousands of writers would like to express but don't because they are always afraid of rejections by a public anchored on the beliefs they have from family roots. They are more than traditional but a path to ride the fears of unknowing, as it is much easier to stay in burning sun than in the shade. People often stay all their life in a job, not because they are happy at it, but because they are afraid of the transition. Others remain living in the same city or country all their life, dreaming of following an ex-friend already gone for better.

As I told my new senior passenger, my writing involves several different personalities, as well as facts and realities, but they are put in a way that will not bore people as the average books do, especially after several pages. People no longer even know what they are reading. This book holds to few personalities, and their names are mentioned often, as the reader could get lost in the world of words. I sometimes mention an important fact a few times, as teachers do with facts that are to be remembered.

Many readers will take weeks to finish the book, and when they begin reading, all the highlights go deep into the memory bank, to be reviewed in the inevitable spiritual world. The bright ones will be even better, as they will feel its importance and just keep reading faster to the last page. For women, I include a few of my designs, and I also put in a few recipes everyone will love. They are easy enough any twelve-year-old could make them. Life is full of variety, or the rainbow wouldn't be in color!

Spirituality is a form of religion, and science is all we need in our material world, but they must work together as perfect partners for us to enjoy life on earth while we are here. The Vatican (Catholics) are trying harder to put religion and science together, as they did with Christ, but it is costing them a millennium of hard work, as lives were lost as martyrs. Finally, Jesus is now an image of love to everyone, even to the ones proudly announcing their belief in nothing. (It's bluffing.) At my daughter's tragic death, a half dozen atheists whispered in my ears that she was now with him. (They didn't say *God*, afraid it would hurt the vanity of superiors.)

The goodwill of the Vatican is notorious, because the leader of all faiths goes to visit the pope, not using the back door, but pompously through the main entrance while CNN is there announcing to us another miracle is happening. The talk, as we all now know, is about love, the same love that put them on a pedestal of solid rock. Jesus died for his principles, which are unchangeable. Facing the cross, there was no option to run from it. He knew that one day, his word would echo around earth eternally, and we can have a better life based on "loving one another," respecting all faiths and borders, holding hands for a life on earth, which is understood to be a preparation for the eternal one of spirit. He, as a spirit, will be there waiting for everyone because the Creator is a just one. Now Christ is revered alongside the Creator. If you pay attention, you will see that from now on, their names are pronounced all the time, everywhere in the good times or at disaster areas or deep in our hearts.

I spent Christmas in Muslim and Buddhist countries. Those religions are in the heart of the general population, but the merry sound of Christmas songs and colorful decorations were everywhere—at the shopping malls, restaurants, and so on. Interestingly, it is in English, to be more realistic, as it all has roots in the good United States. Even in Brazil, as I listened to Christina's songs, I noticed something interesting: all the children on the chorus at the mall or any public appearance were singing in English. It gave me a feeling it was another Christmas miracle by him.

While the senior was analyzing the book, I asked him about his religion. He told me it was irrelevant, as it is private in people's hearts; we all have one God as our Creator and perfect laws for all of us in the universe. The eternal life as spirit is part of our existence as intelligent human beings. I told him learning is an eternal path here as it would be there in spirit. He just gave a great answer related to the religious status of each one of us. He said in a few months, he would check Amazon to see if the book is already there. He said if it depended on him, this book would be a best seller. Before I could say anything, his wife came and dragged him away to meet someone. We said, "I'll see you later."

After ten days of reading, writing, talking to anyone I could from the more than three thousands souls, enjoying good food and nightly family shows, while watching the vessel continually cutting the small waves with its massive weight, not going against gravity, but negotiating with it to remain afloat, I felt happy, but at the same time, my mind was on the billions of souls suffering on our planet and those with intellectual and financial difficulties, being on the same boat, being in a world of uncertainty, as we are all asking questions without answers. Some of us desperately claim to have them, as they are in quicksand up to their noses, while they bring anarchy where there should be love, the dove of hope.

This book came from deep within my soul. I am getting ready to cruise to other dimensions, but my spirit will be here, doing what I can, as others are on the same path. We can't stop, because this way, there will be no uninteresting times, as souls or spirits.

Love to all,
William Moreira (Canno)
Grand Holiday (Ibero Cruisers)
March 3, 2013, upon arrival in Rio, just off Brazil's coast

24 A Personal Prayer to God

God, our Celestial Father, you are heaven almighty and kind, heaven giving strength to those who are going through a difficult endeavour, giving light to those who seek the truth. Place into the hearts of men compassion and charity.

God! Give to the traveler a guiding star; to the afflicted, consolation; to the ill, peace.

God! Give to the guilty, regret; to the spirit, the truth; to the child, guidance; to the orphan, a family.

Lord! May your mercy extend itself over everything you have created.

Give mercy, O Lord, to those who do not know you, hope and understanding to those who suffer.

May your kindness permit compassionate spirits to spread peace, hope, and faith over everyone.

God! One stroke of lightning, one spark of your love can flare the earth. Let us drink from the fountain of your abundance, your infinite kindness, and all tears will be dried, all pain alleviated.

Only one heart, only one thought, will elevate us to you—recognition of love. Like Moses on the mountain, we wait for you with open arms.

O Power! O Kindness! O Beauty! O Perfection! We want, in some way, to reach your mercy from our wrongness.

God! Give us heartfelt charity; give us faith and understanding; give us the simplicity that will make our souls the mirror upon which your image can be reflected.

This prayer is our beginning in the direction of our contact with the spiritual world and absolute love; the love of our Creator is waiting for us all, in our free will, to reach his arms.

25 The Message of the Millennium

Lord, my God!

I have walked lands, navigated the seas, flown over horizons, conquered forests, climbed mountains, and worked to win with the sweat of my brow for our daily bread. I have studied, researched, learned, taught, and suffered in order to feel the meaning of "love one another" and to learn why life is like it is.

Now, my Lord, my eyes are tired, my legs are too weak to take long walks, my skin is dry with the wrinkles of old age, and my vocal cords are unable through words to rescue your children, my brothers.

Above my white head appears a smiling lad—happy, almost angelic, floating clearly in front of me, as part of my thoughts.

Then, in contentment, confronting a radiant positive spirituality, I ask our Lord, "Who is this fascinating lad?"

And the Lord answers, "This lad is you, reflected in the mirror of truth, showing what you offered in the name of love and charity. This fluid image of energy is your eternal spirit that never tires, ages, or dies. What you gave as a seedling, you are harvesting, and now you belong to the celestial world, where heaven has no limits."

Moses's Ten Commandments and the Eleventh Commandment

Jesus's *parables* are spiritual guidance for everyone, as are the moral laws of progress and Moses's Ten Commandments written millennia ago. Now, I have spiritually received an amendment necessary as we change our behavior in society.

1. "Thou shalt have no other gods before me." (More than one God would be no God at all.)

2. "Thou shalt not make unto thee any graven image." (God has no portrait to be adored because he is in our conscience.)

3. "Thou shalt not take the name of the Lord thy God in vain." (Do not swear lies; tell the truth in his name.)

4. "Remember the Sabbath day, to keep it holy." (Everyone deserves at least one day off after working six days in the week, even slaves in the past.)

5. "Honour thy father and thy mother." (When parents get older and have no more physical protection or cannot support themselves, many sons and daughters abuse them as a lack of love.)

6. "Thou shalt not kill." (If this was in everyone's heart, there would be no wars or any assassination, but the only killing allowed would in self-defense when our lives are in peril.)

7. "Thou shalt not commit adultery." (This is wrongly interpreted as being only for men, but it is as also for women.)

8. "Thou shalt not steal."

9. "Thou shalt not bear false witness against thy neighbor." (It is very serious to lie against anyone because witnesses can destroy someone's life as a wrong not committed.)

10. "Thou shalt not covet." (This is to crave or desire something another has and could be explained as jealousy of someone with success in any field.)

11. "Thou shalt not desire another man's woman or another woman's man." (The reason is because today equal rights laws, as moral progress for women, give them rights in male-dominated society, and some women are using their freedom to cover immoral sexual behavior, which is destroying the blessed concept of family, bringing anarchy to moral law affecting the spouse, children, parents, and friends without any repair sometimes for generations.)

If you have any comments, please feel free to send me an e-mail, and I will do my best to answer: williamcanno@gmail.com.

Designs for Women by the Author

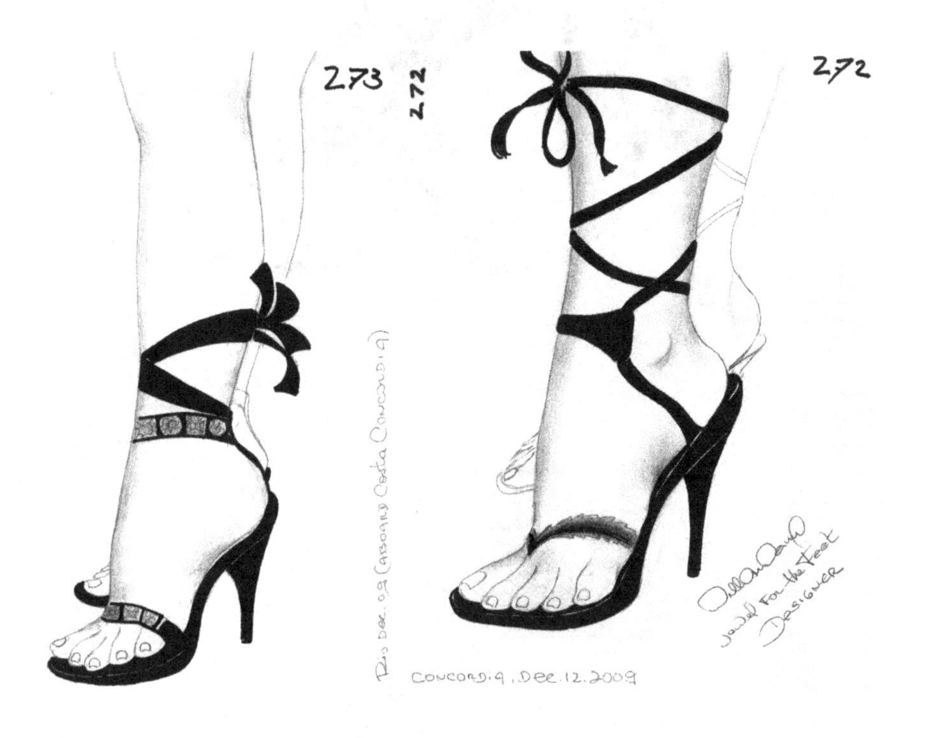

Rio Dec 08

NYC AUG 08

126

Rio/July/08

262

CONCORDIA, DEC.09 (Costa)

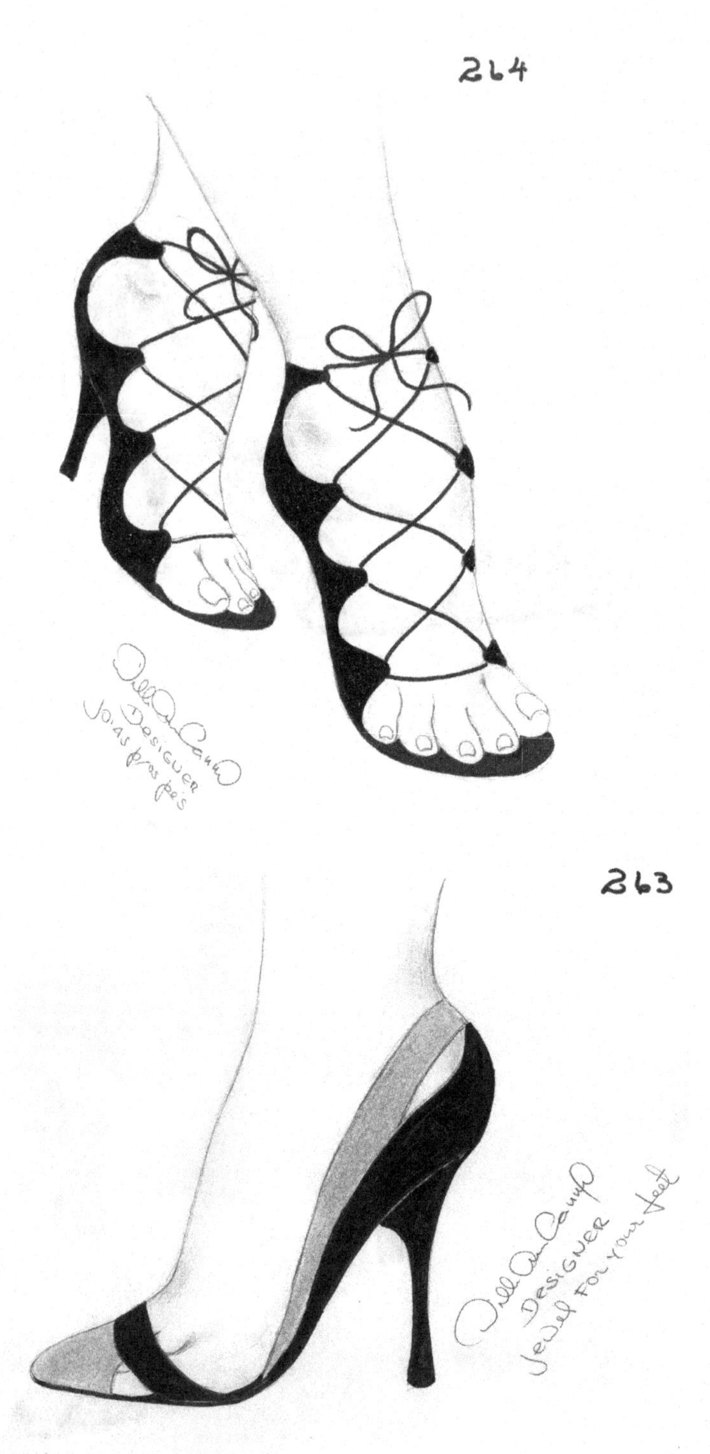
264
DESIGNER
Joias pros pés
263
DESIGNER
Jewel for your feet
CONCORDIA, DEC.09 (COSTA)

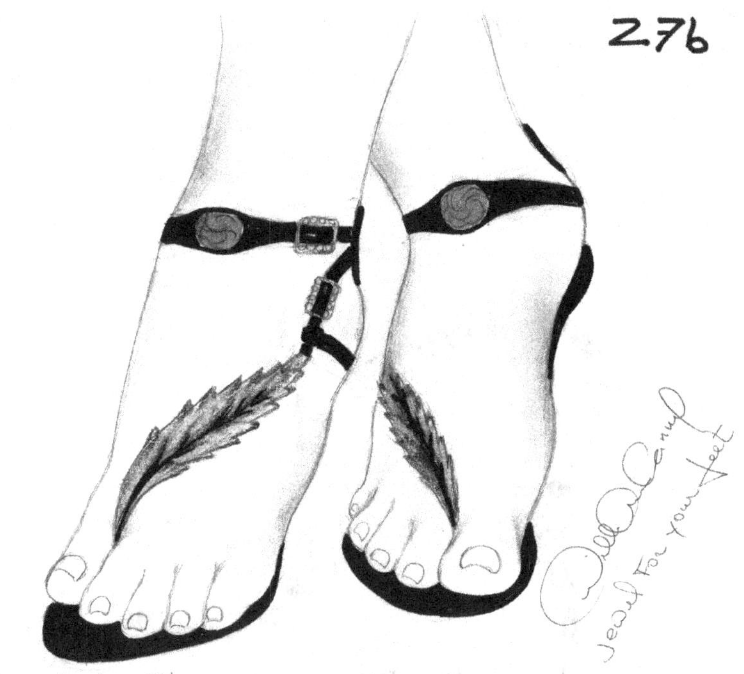
276
Jewel for your feet
COSTA CONCORDIA - RIO - DEC. 09

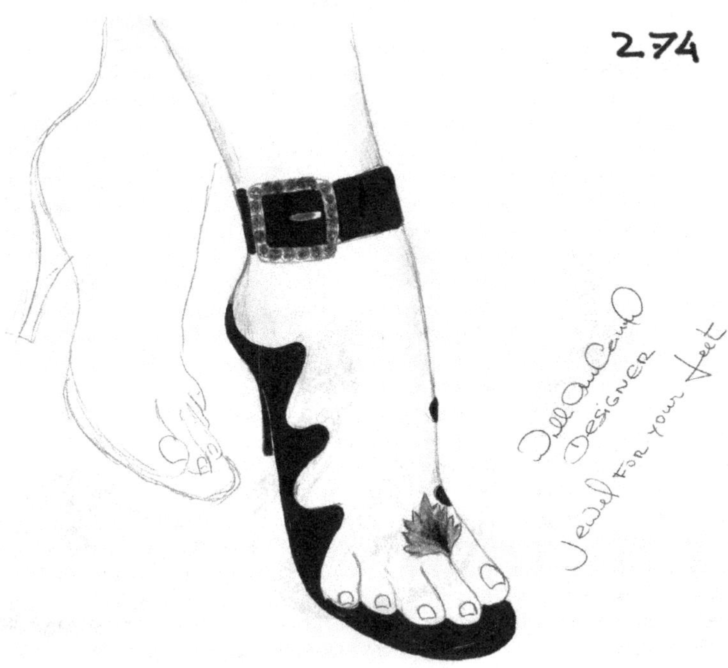
274
DESIGNER
Jewel for your feet
DEC. 09 : RIO (CONCORDIA - COSTA)

234

NEW YORK, Jan 08

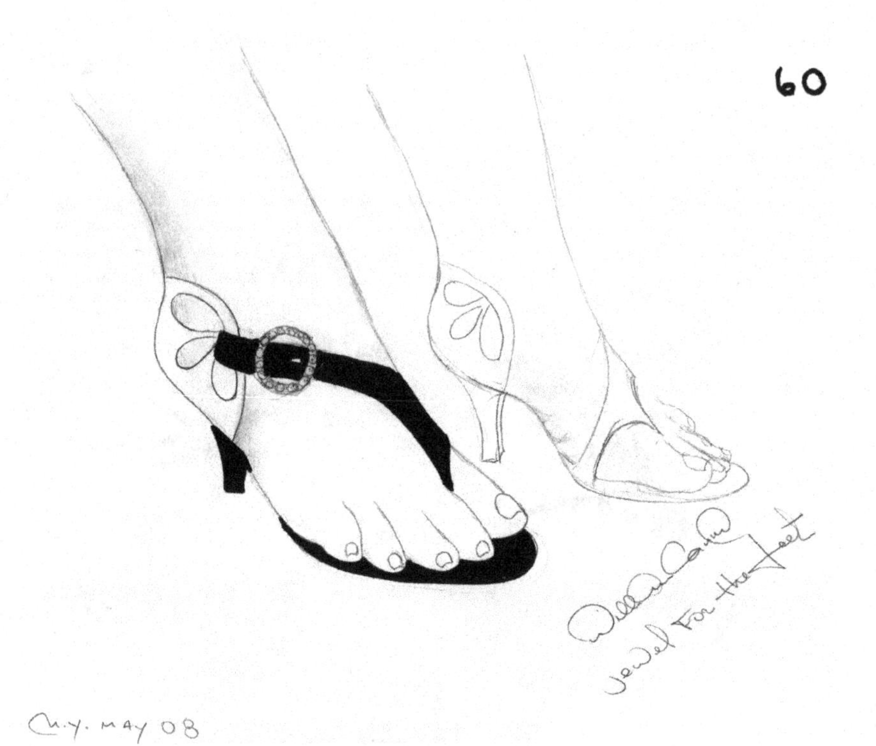
60
Jewel For the Feet

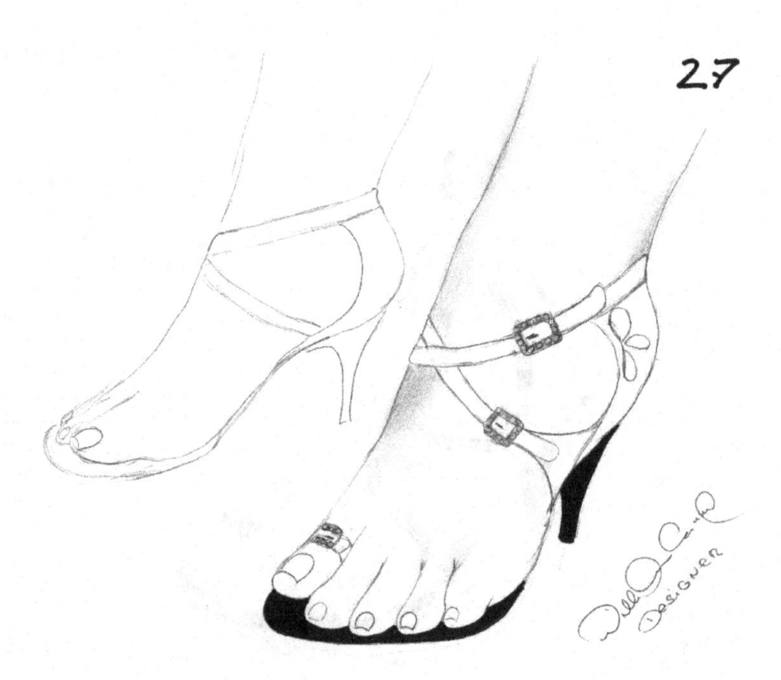
27
DESIGNER
Rio Feb. 08

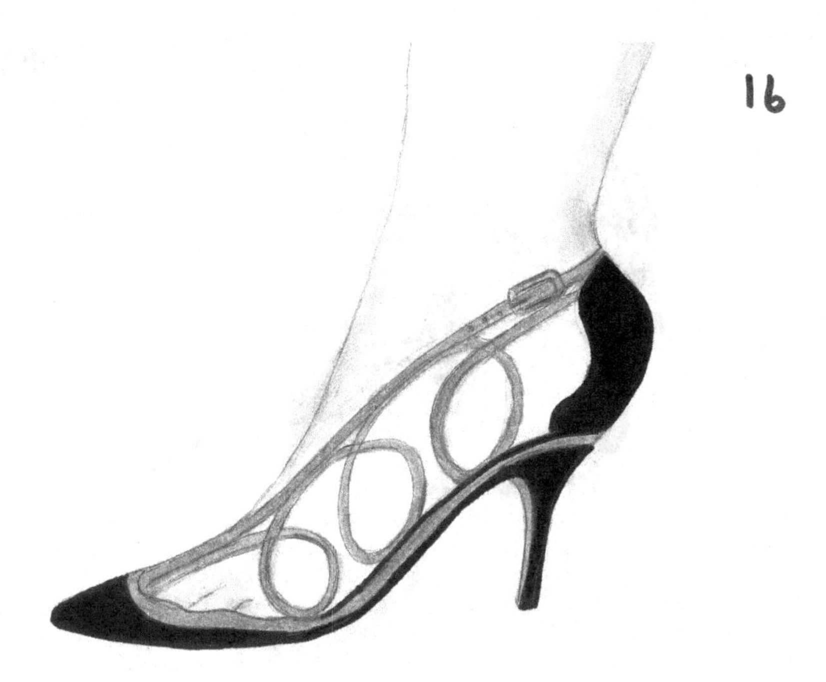
16

NYC/Jan/08

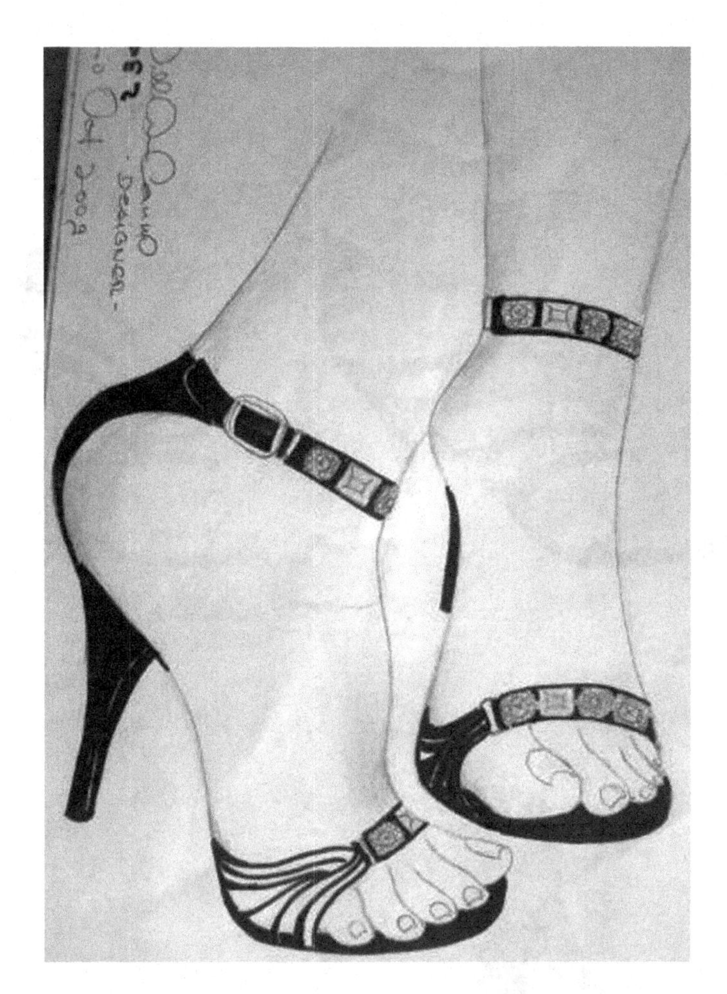

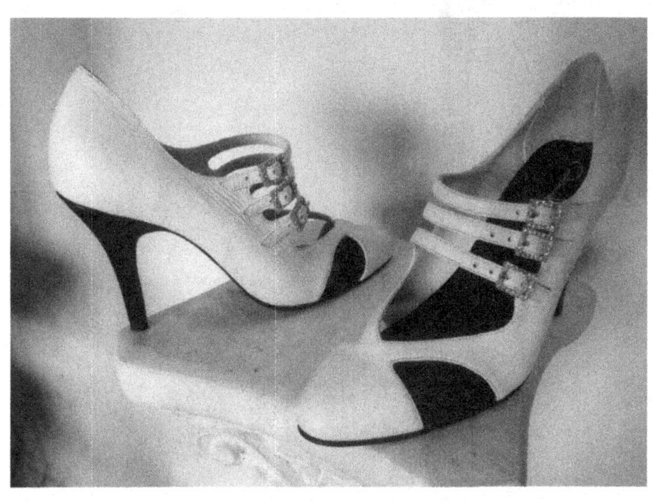

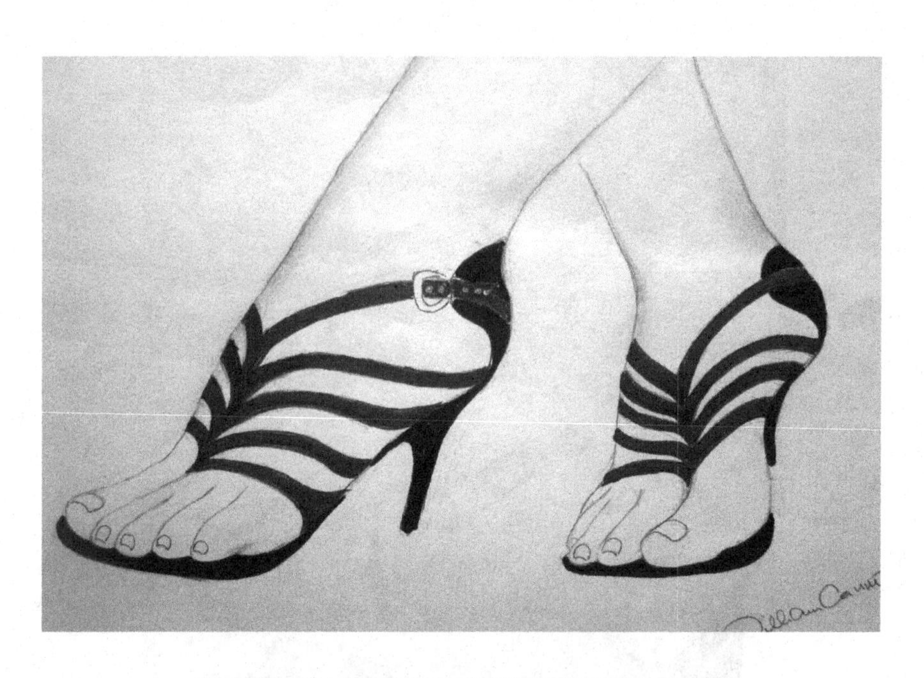

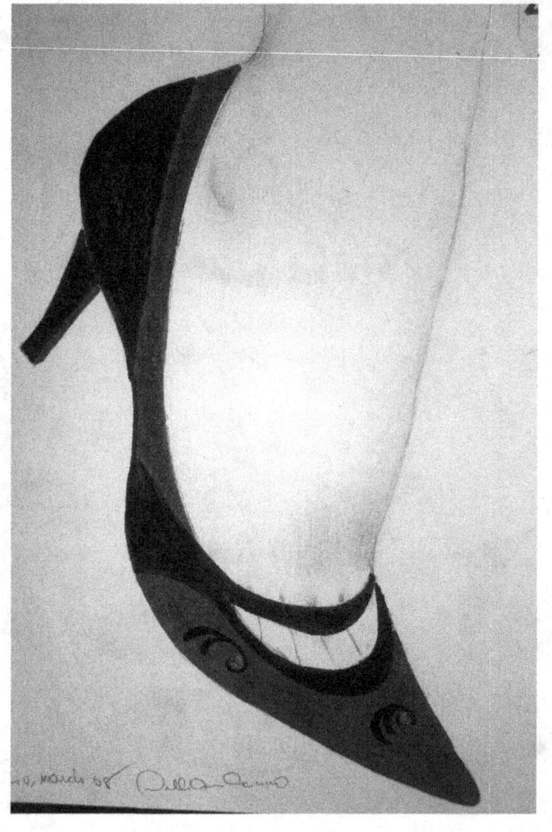

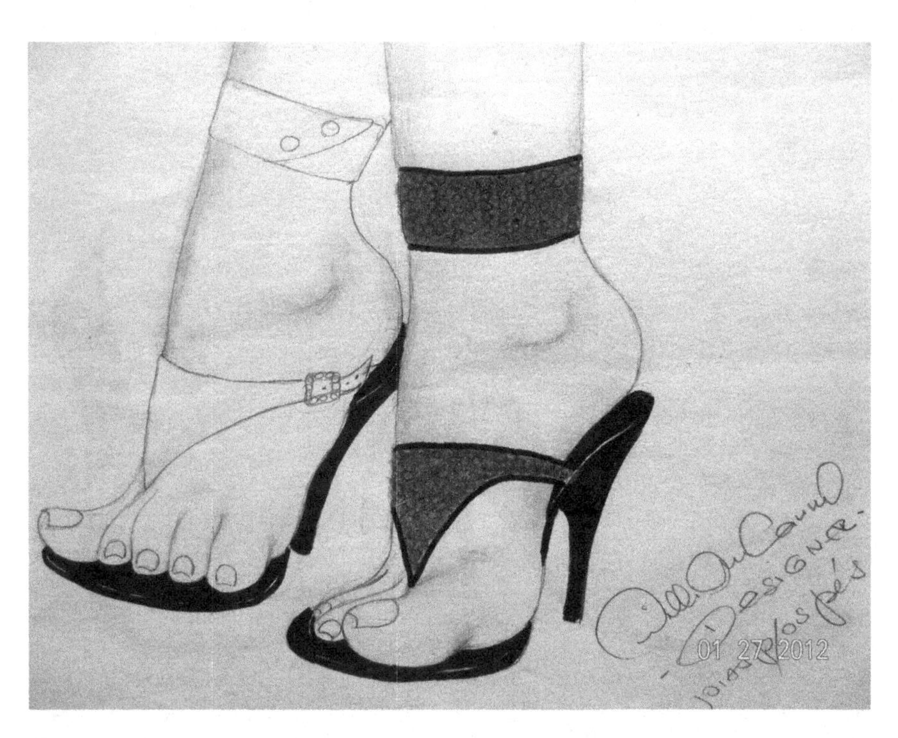

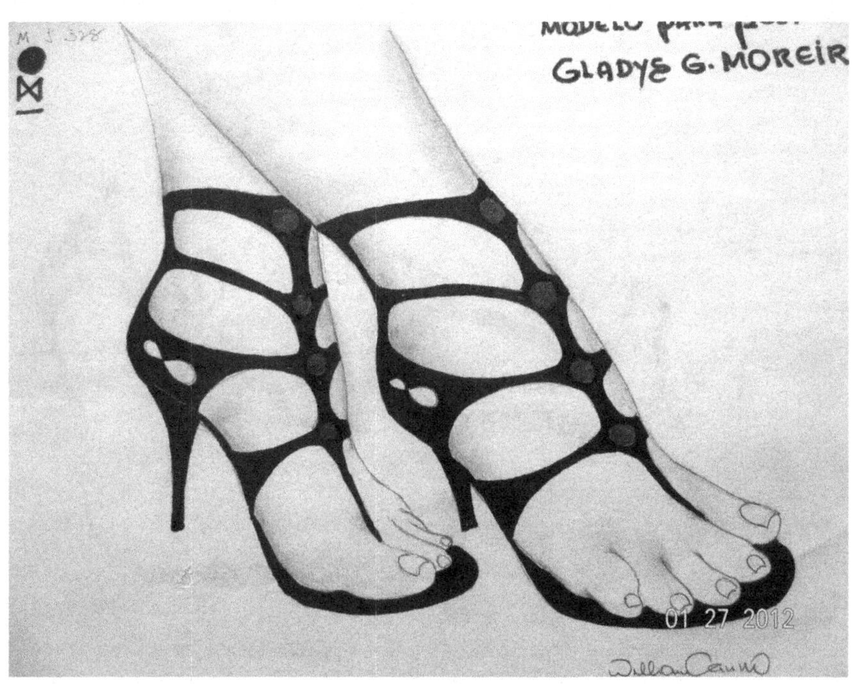
GLADYS G. MOREIR
01 27 2012

01 27 2012

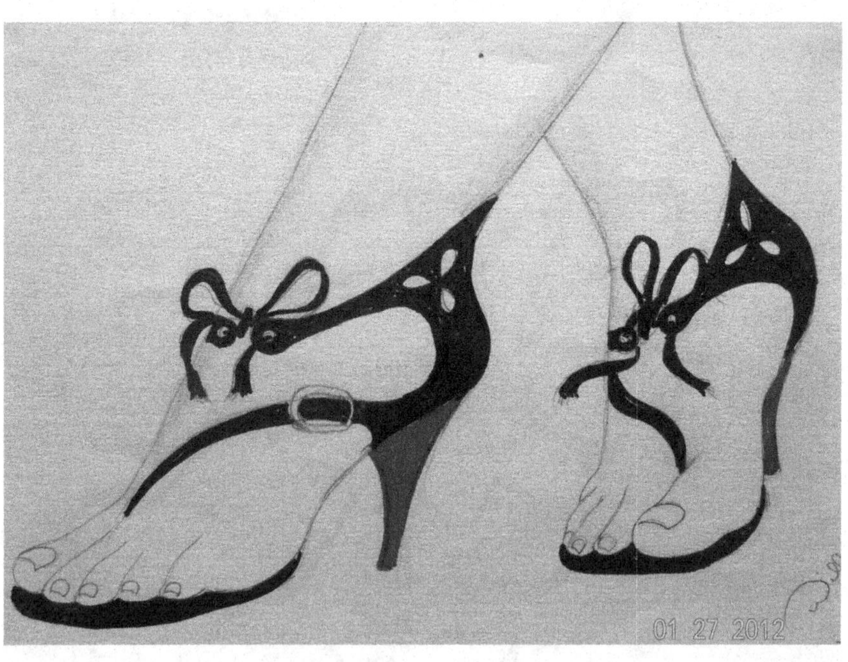
01 27 2012

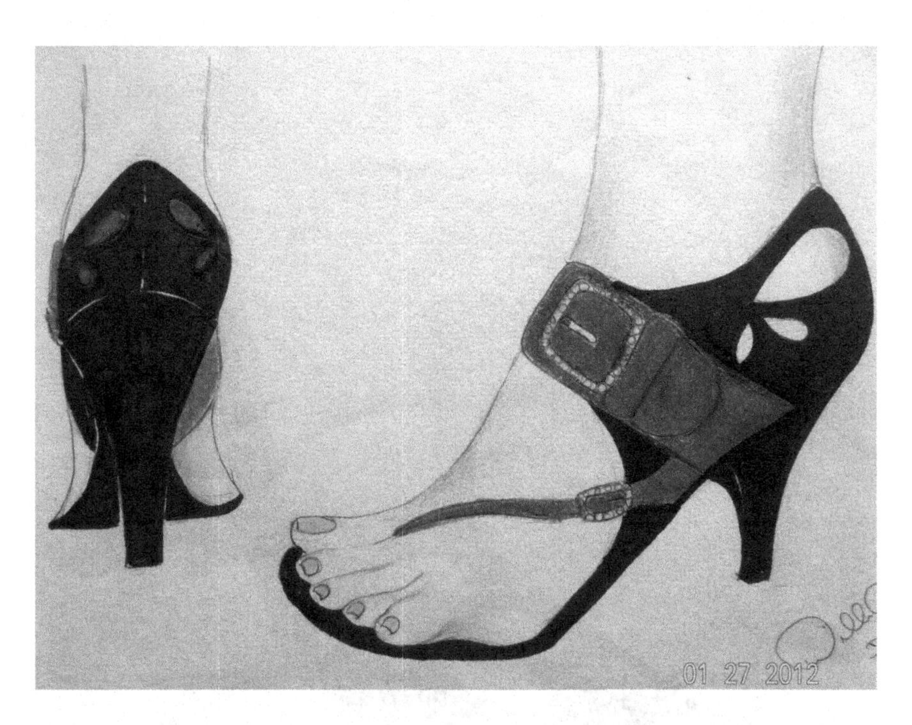
01 27 2012

01 27 2012

01 27 2012

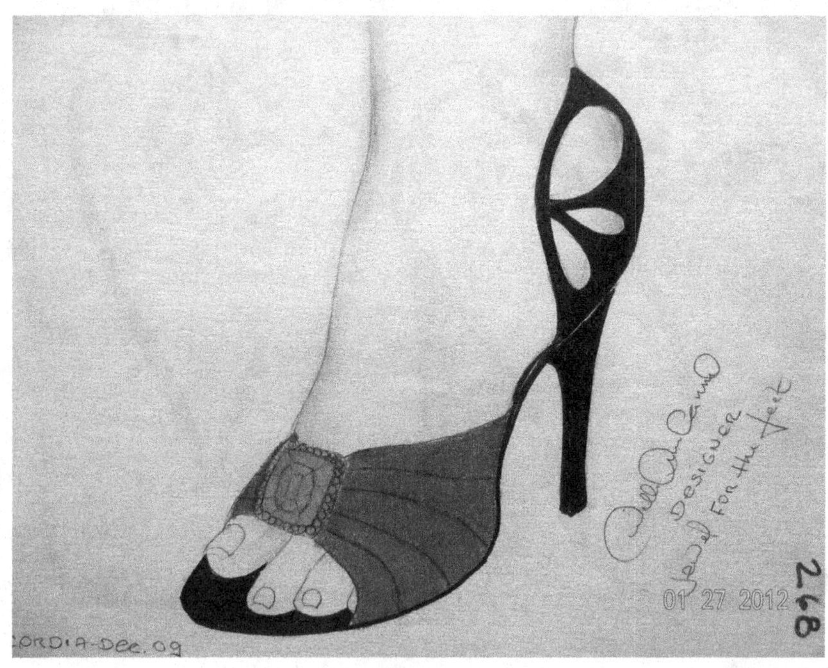
ORDIA-Dec.09
DESIGNER
For the
01 27 2012
268

01 27 2012

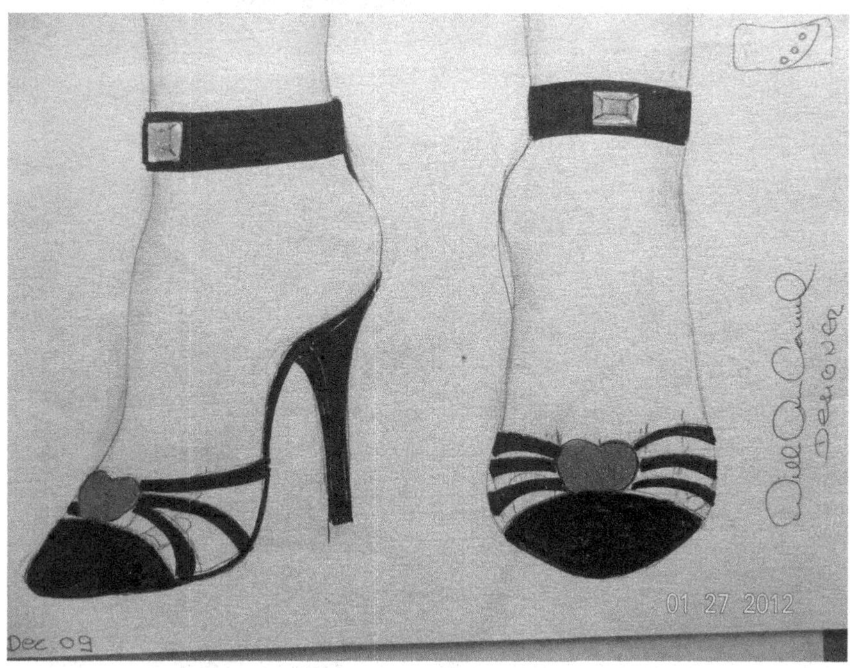
DESIGNER
01 27 2012
Dec. 09

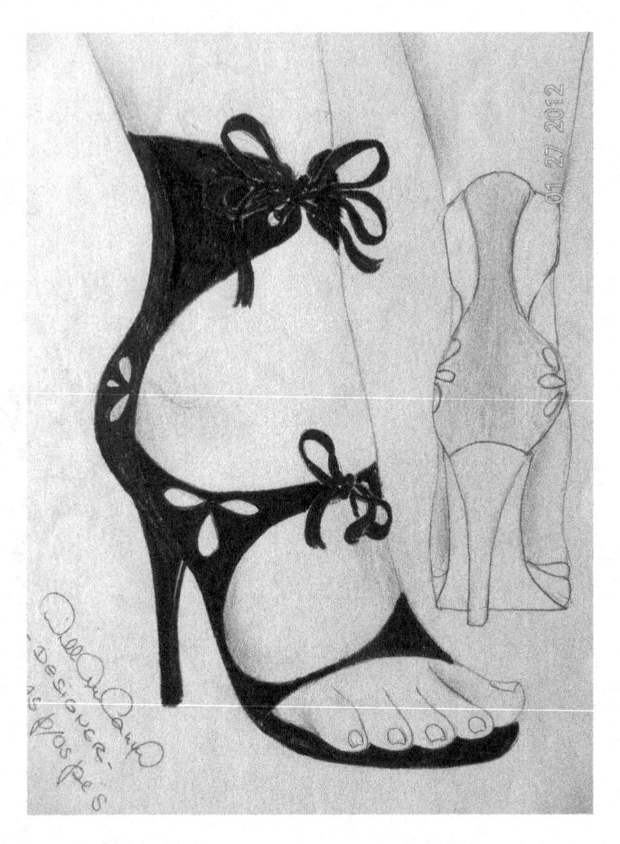
01 27 2012
-DESIGNER-

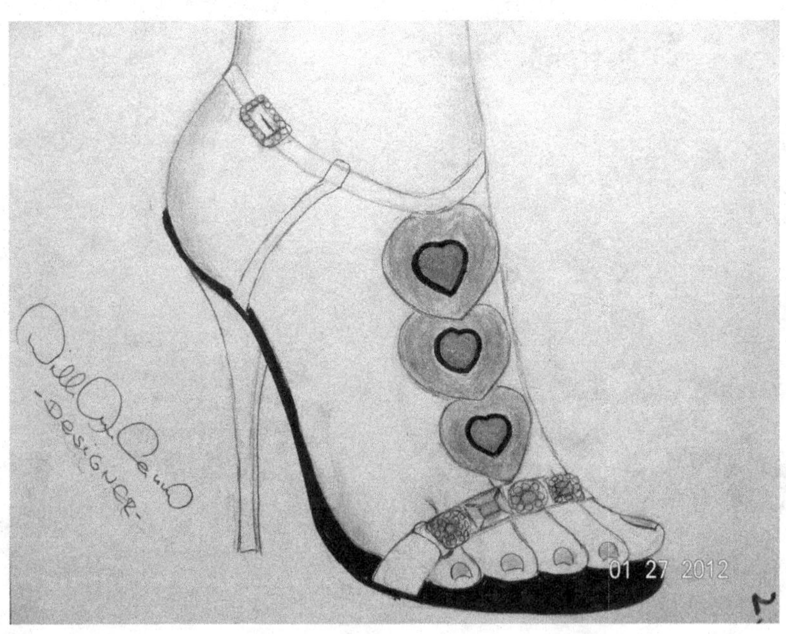
-DESIGNER-
01 27 2012

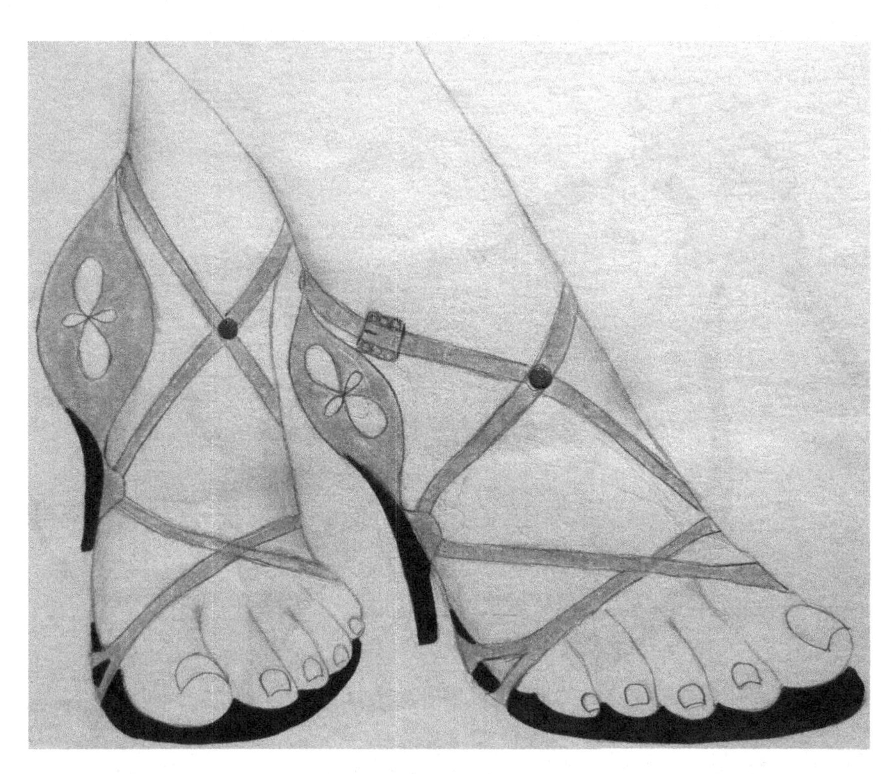

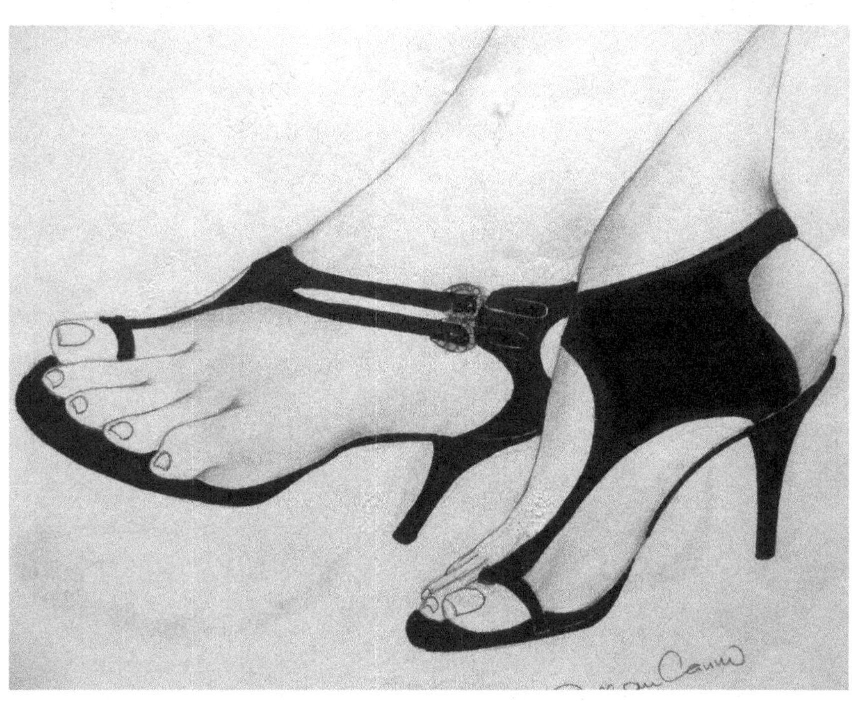

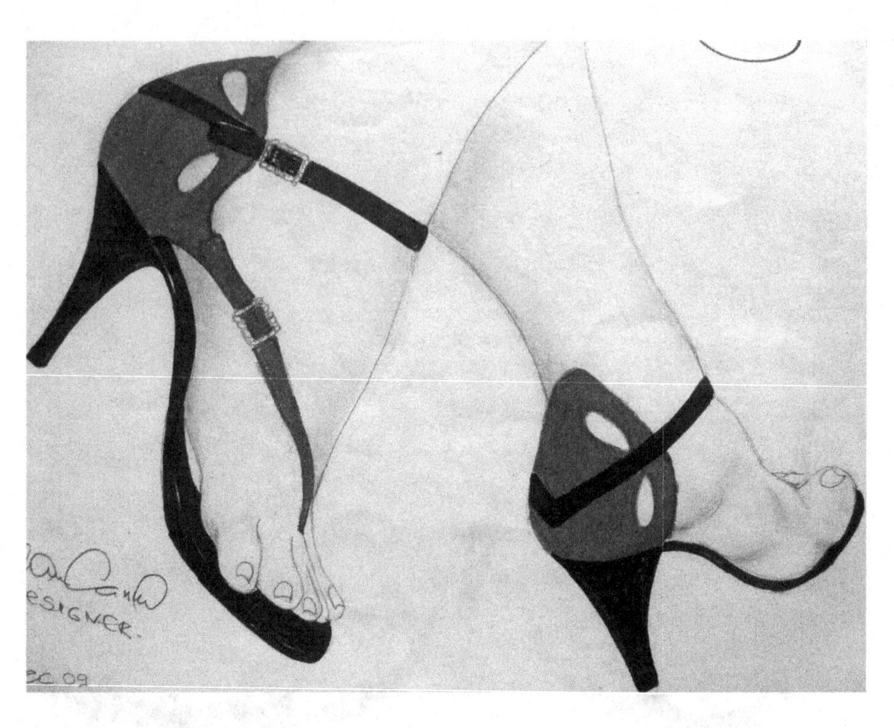

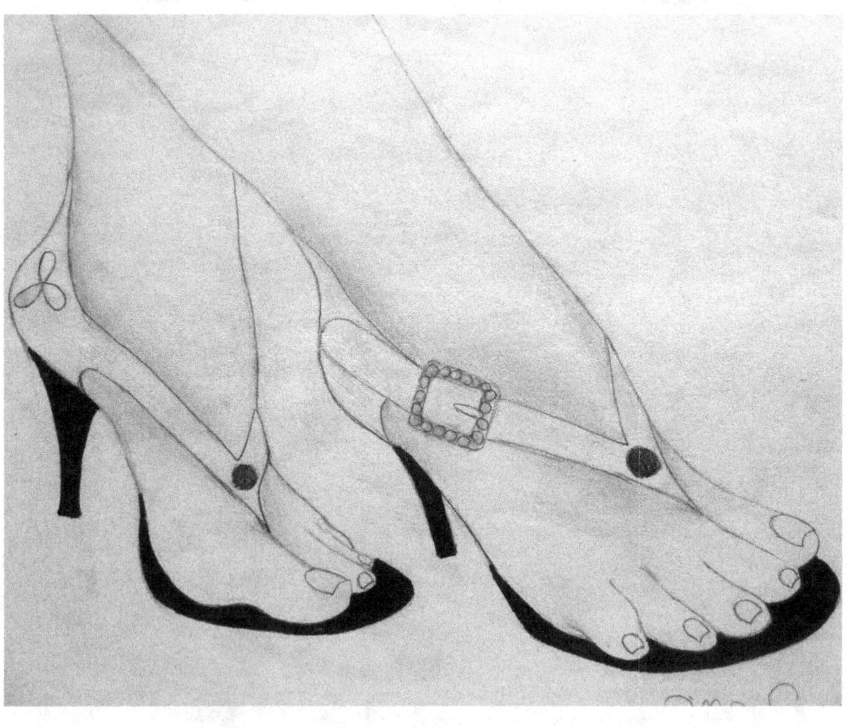

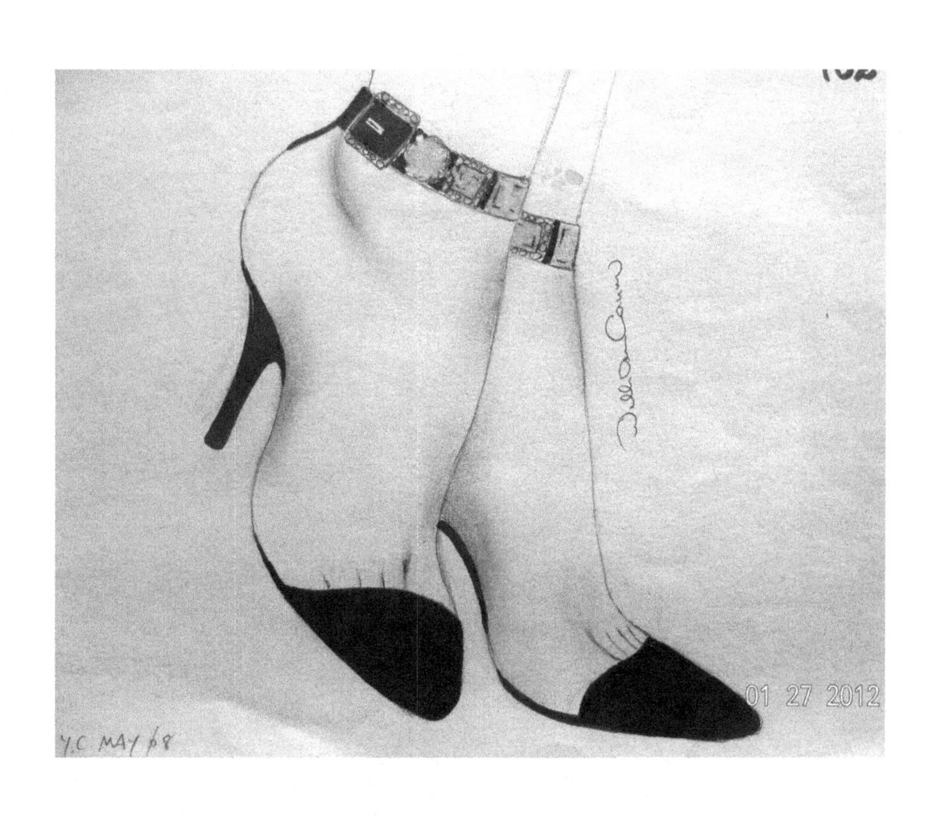
01 27 2012
Y.C MAY 08

Oil Paintings by the Author

Four steps for an easy **POTATO PANCAKE** as everyone will love it for breakfast, lunch, dinner or in the wee hours. (I believe is no need to write it down).

One large **POTATO** (peeled)

One **ONION** (half size the POTATO)

Grade BOTH on the large side of the hand grader on the top of a clean KITCHEN TOWER. Dry it twisting the towel of all the LIQUID from the potato (S) and onion (S).

PUT IN A BOWL and add:
2 – **EGGS** and 1 – Soup spoon of **Flour**.

MIX THE 4 INGREDIENTS WITH A FORK (I use my fingers).

NOTE: Salt, black or white pepper, chopped sweet red or green pepper, parsley, small amount of cheese, bananas is your option to glorify it.

SAUTÉ (sounds more classical than fry) in a large pan (use what you have) on a thin coat of HOT OIL (do not use olive-oil) it will deflect the flavor. It is a flat pancake and it will take only few minutes to be crisp outside and moist inside - your success will be absolute.

NOTE: For more PANCAKES just multiply the ingredients as 2 and 2 equals 4.

SERVE WARM with SOUR CREAM, APPLE SAUCE and CREAM CHEESE

"To make it easier to remember: POTATO, ONION, EGGS and FLOUR"

My daughter ***CAROL*** in kindergarten could do this ***ONE STEP APPLE SAUCE:***

8 to 10 SMALL HARD APPLES: Peel it cut in pieces, put in a ONE QUART POT, add:

I soup spoon of **SUGAR** and 1 spoon of WATER. Cover it and cook in medium fire until tender (20 minutes). Wisp with a HAND ELECTRICAL MIXER. Last one week in the freeze.

TO EAT WELL IS PART OF BEING ON EARTH AS A GIFT from HEAVEN.

Observation: I was 8 years old when my mother taught me this recipe as I honor it from the ONE BILLION IN MY MIND because recipes are endless, as we can mix ingredients, fry, deep fry, sauté, boil, broil, grill, smoke, salty, sweet, dry, eat raw and marinated the thousands of elements the merciful Creator have intelligence and time to give to us all, but we have to deserve it in our good deeds.